THE PSW SCIENCE BULLETIN

Founded 1871 as the Bulletin of the Philosophical Society of Washington

2,500TH MEETING ★ SEPTEMBER 6, 2024 ★ SCIENTIA GRATIA SCIENTIAE

VOLUME 17

First Edition

Commemorating the Society's
2,500th Meeting

6 September 2024

The PSW Science Bulletin: Volume 17

Westphalia Press
An imprint of Policy Studies Organization
1527 New Hampshire Ave., NW
Washington, D.C. 20036
info@ipsonet.org

ISBN-13: 978-1-63723-656-7

Daniel Gutierrez-Sandoval, Executive Director
PSO and Westphalia Press

Updated material and comments on this edition
can be found at the PSW site:
pswscience.org

The PSW Science Bulletin
Volume 17

Founded 1871 as the
Philosophical Society of Washington

Edited by
Larry S. Millstein, PhD, JD
President
and Robin M. Taylor, PhD, JD
Corresponding Secretary

1st Edition
August 2024
www.pswscience.org

WESTPHALIA PRESS
An imprint of Policy Studies Organization

TABLE OF CONTENTS

Forward

Larry S. Millstein, PhD, JD
President
PSW Science
Founded 1871 as the Philosophical Society of Washington

The last volume of the *Bulletin of the Philosophical Society of Washington*, Volume 16, was published in 1962, now 62 years ago. And that volume itself was the first since Volume 15 was published in 1913, 51 years earlier.

So why the current publication? The immediate impetus for the current volume, designated Volume 17, 1st Edition, is the occasion of the Society's 2,500th meeting, to take place on September 6, 2024 in the Warne Ballroom and Powell Auditorium of the Cosmos Club in Washington, DC. It seemed fitting in commemoration of the event to bring together in one publication some of the important documents about the Society's formation and history.

Towards that end you will find President Joseph Henry's address explaining how and why the name was arrived at, why the Society was formed and its purposes, the proper conduct of discourse at the Society's meetings (and all scientific discussions), several histories of the Society, a memorial tribute summarizing the life and contributions of Henry (founder and first President of the Society), a list of Nobel Laureates who have given PSW Science lectures, a list of the Joseph Henry speakers, and the minutes of several early meetings at which important results were disclosed paired with more recent PSW Science lectures relating to the same topics.

Looking over this collection, and perusing the sixteen volumes of the *Bulletin*, reveals an organization continuously engaged for over 150 years with the scientific and important knowledge creation activities of the day. Year after year speakers have come to the Society's meetings to engage with members and present their results and views on just about every topic of scientific importance, spanning an unusually wide breadth of disciplines – mathematics, the physical sciences, the life sciences, anthropology, archeology, linguistics, and economics (among others). And, typically, they have been among the most prominent experts on the topics of their discourse. More than 24 Nobel Prize recipients have given lectures at PSW Science, and 22 of the annual Joseph Henry Lectures given so far have been given by Nobel Laureates.

It is hard not to be favorably impressed by the breadth of topics addressed at the Society during its more than 150 years of operation, and it is noteworthy that many topics that were of interest in the past continue to be of importance and attention today. The Society was one of the first venues to host Gray's and Bell's demonstration of their respective telephones and to hear the debates between Bell and Gallaudet on the proper attention to be paid to the deaf (a debate that still echoes today). It was among the first venues where John Tyndall presented his measurements of the heat capacities of atmospheric gases (so important to climate science today). Albert King presented compelling evidence mosquitos transmit malaria, 14 years before it was proven experimentally. More recently the Society hosted an early presentation on a gene editing approach to prevent the malaria parasite from reproducing in its fly vector.

The Society has maintained its traditions throughout its history, with scant modification, while nonetheless adapting to rapidly changing social environments. The Society still has as a crucial part of its program in-person experiences approximately every two weeks, always including a pre-lecture social reception and dinner for members, the speaker(s) and their guests, a focal lecture open to the public by a prominent expert on a topic of current scientific importance, and a closing social hour for discussion and interaction among members and guests with speakers. It continues to publish minutes of its meetings, currently on the PSW Science website (instead of hard copy paper publications), and in multimedia forms including graphics and video as well as text.

The Society in recent years has expanded its activities for the increase and diffusion of scientific knowledge by utilizing new media modalities unknown to the founders. Viewership of the Society's lectures is preponderantly via the internet. On average 20 to 30% (or more) of live participation in the lectures is via the live stream on the

PSW Science YouTube Channel. Sometimes the live stream audience far exceeds the auditorium's capacity. Recordings of the lectures posted to the PSW Science YouTube Channel are viewed by a far larger audience, averaging over 1,000 views for each video. Some videos have been viewed several hundred thousand times, and the roughly 225 videos posted so far have been viewed, cumulatively, over 1,000,000 times. The on-line audience now exceeds by orders of magnitude what is possible for in-person attendance, and the trend is certain to continue and perhaps accelerate with time. And it is worth noting that only by utilizing Zoom was the Society able to maintain and even exceed its normal cadence of meetings on-line through the entire COVID-19 pandemic! Media modalities will be an increasingly important aspect of the Society's activities in fulfilling its missions of the increase and diffusion of scientific knowledge.

As the Society embarks on the next 1,000 meetings it will continue its tradition of in-person gatherings and now on-line offerings featuring interesting and challenging speakers and topics, provocative and constructive discussions, camaraderie, and success in bringing science understanding to as wide an audience as possible throughout the world.

I hope you will enjoy looking through this first edition of Volume 17 of the *Bulletin* and that it will stimulate appreciation of the Society's rich intellectual traditions and the ways in which the Society can continue to support science, seriously.

Larry S. Millstein, PhD, JD
President and Program Chair
PSW Science
Founded 1871 as the Philosophical Society of Washington
August 5, 2024

2500th Meeting Lecture
September 6, 2024

"Uniting All the Men of Science in this City in One Organization": Joseph Henry, the Philosophical Society of Washington, and the Status of Science in Post-Bellum Washington

Marc Rothenberg
Editor, The Joseph Henry Papers Project (Ret.)
Research Associate, Smithsonian Institution Libraries and Archives

"Uniting All the Men of Science in this City in One Organization": Joseph Henry, the Philosophical Society of Washington, and the Status of Science in Post-Bellum Washington

Marc Rothenberg

Editor, The Joseph Henry Papers Project (Ret.)
Research Associate
Smithsonian Institution Libraries and Archives

If you know the musical "Hamilton," you know it opens with Aaron Burr asking a question:

> How does a bastard, orphan, son of a whore and a
> Scotsman, dropped in the middle of a forgotten
> Spot in the Caribbean by providence, impoverished, in squalor,
> Grow up to be a hero and a scholar?[1]

I am going to open with a similar question: How did the son of an alcoholic day-laborer, with minimal formal education, help turn Washington into a major center of scientific activity, become the symbol of American science in the last third of the nineteenth-century, and along the way, become the driving force behind the founding of the Philosophical Society of Washington (hereafter PSW)?[2] I want to talk about Joseph Henry, the most important nineteenth-century American scientist most people have never heard of.

On May 13, 1878, Joseph Henry, Secretary of the Smithsonian Institution, President of the National Academy of Sciences, President of the PSW, and among other achievements, first science advisor to a President of the United States, died. Three days later, his funeral was held at New York Avenue Presbyterian Church, The federal government closed for the funeral, which was attended by President Rutherford B. Hayes, Vice-President William Wheeler, the Cabinet, members of the Supreme Court, members of Congress, senior officers of the Army and Navy, and representatives of scientific societies, and academia (including members of the PSW).[3]

Commemorations of his life soon followed. Among these was a meeting of the PSW on October 26, 1878, a memorial service in the House of Representatives chamber on January 16, 1879, and statuary. Speaker after speaker emphasized Henry's role in public and government service. At the PSW meeting W. B. Taylor claimed: "Whenever in any important case a scientific adviser could be useful to the proper conduct of a Bureau, Henry's reputation generally pointed him out as the most suitable expert and arbiter.[4] At the memorial in the Capitol, Congressman (and future President) James A. Garfield told his audience that under Henry the Smithsonian was "one spot in this city across which the shadow of partisan politics has never fallen." Henry was "a man who served his Government with singular ability and faithfulness."[5] But General William T. Sherman made an even stronger claim for Henry. To him, Henry was a "hero":

> "His conquest was not over cities razed , homes desolated, or the forms crushed and lacerated, but over the obstacles of nature, in mastering her laws and harnessing them to the uses of his fellow-men. No widows or orphans are left to mourn over his victories, but millions who have reason to rejoice in the increased knowledge and stimulated industry which followed in the wake of his intellectual triumphs."[6]

This scientist was "an example to the youth of our land, to show that honor and fame may be earned in the school of philosophy as well as in the more tempting and active scenes of public life."[7] Sherman's description of the scientist as conquering hero without resulting death or destruction is perhaps all the more powerful because of his own reputation as a ruthless warrior.

In 1883 the nation showed its esteem for Henry in another way. On April 19 the federal government closed early and up to 10,000 spectators, including top government officials, crowded the Mall near the Smithsonian Building, to hear John Philip Sousa lead the Marine Band in a new composition, "The Transit of Venus March," written for the occasion and to see the unveiling of the statue of Henry we are all familiar with, now in front of the Smithsonian Castle.

International recognition followed in 1893 at the International Congress of Electricians, where the international units of electrical measure were named. Along with the volt, the amp, the farad, the coulomb, the joule, and the ampere, there was the *henry*, the unit of inductance.

But the ultimate acknowledgement of Henry's status in the minds of late nineteenth-century Americans can be seen in the Main Reading Room of the Library of Congress Building (completed in 1897). If you look up to the balustrade you will sixteen statues representing eight areas of human endeavor and accomplishment. The two men selected to represent science are Sir Isaac Newton and Joseph Henry. They join, among others, Homer, Beethoven, Shakespeare, Plato, and Michelangelo. Not bad company and a measure of what Henry meant to nineteenth-century American science and to the nation.

So who was the man Sherman called a hero?

Joseph Henry[8] was born in Albany, New York, on December 17, 1797, according to baptism records.[9] Henry himself always asserted later in life that he was born in 1799. Making himself two years younger may have been linked to possible sensitivity as to the manner of his education, which I will touch upon below.

Henry's father was an alcoholic, doomed to die in a fit of delirium tremens in 1811, when Henry was thirteen. At the time Henry was living in the village of Galway, New York, about 37 miles from his parents in Albany. He had been sent to Galway at about age eight, to be raised by his step-grandmother and an uncle, probably due to the limited financial means of his family. Galway offered Henry a village school education, where he learned the fundamentals: arithmetic, reading, and writing (but very little spelling, as his future letters would show). He did not return to Albany to live with his widowed mother, now running a boarding house, until 1814 or 1815, and there became apprenticed to a silversmith (not very successfully). Having fallen in love with the theater during visits to Albany while still living in Galway, on his return to Albany he also considered a career as an actor.

It was at this point in his life that Henry, the unhappy apprentice, with little money and an uncertain future, found a new passion which changed his fortunes. It was science, and the catalyst for his transformation was the first volume of George Gregory's 1808 book *Lectures on Experimental Philosophy, Astronomy and Chemistry.*[10] (The book belonged to one of Mrs. Henry's borders struck by Henry's response to the volume, the border gave the book to Henry.) Gregory's popular introduction to the physical sciences begins with an appeal to the readers' curiosity and desire for social advancement: "You are, I presume, desirous, my young friends, of acquiring knowledge, of satisfying your curiosity, of storing your minds with useful ideas, of fitting yourselves for company and conversation, and of enabling yourselves to proceed gradually in the paths of science, till you arrive at distinction and eminence."[11] The book goes on to answer a long series of questions about natural phenomena. As indicated by Henry's marginal annotation, one set of questions in particular attracted his attention: You throw a stone, or shoot an arrow upwards into the air; Why does it not go forward in the line or direction that you gave it? Why does it stop at a certain distance, and then return back to you? What force is it that presses it down to earth again, instead of its going onward?"[12] This passage contains, of course, some of the most basic questions in physics.

There is no need to guess Henry's reaction to Gregory's book. He summarized the book's impact in an inscription written in 1837 upon the presentation of his copy to his son William, then five:

> This book although no means a profound work has under Providence exerted a remarkable influence on my life. It accidentally fell into my hands when I was about 16 years old and was the first book I ever read with attention.
>
> It opened to me a new world of thought and enjoyment, invested things before almost unnoticed with the highest interest, fixed my mind on the study of nature and caused me to

resolve at the time of the reading it that I would immediately commence to devote my life to the acquisition of knowledge.

Despite these noble sentiments, the reality remained that Henry was a fortuneless young man who had to toil for a living. He worked as an actor, a country schoolteacher, a private tutor, and a road surveyor. He interspersed his work with part-time attendance at the Albany Academy between 1819 and 1822, although he was almost a decade older than his classmates (although the head of the Academy apparently thought Henry was only sixteen at the time of his enrollment). Catering to boys typically from the middle- or upper-class, the school paid special attention to mathematics and physical science. The connection with the Albany Academy led to his appointment in 1826 to the Academy's faculty and the true launching of his scientific career. In 1830 he married his first cousin, Harriet Alexander, whose father was a successful businessman and had left extensive land holdings upon his death in 1809.[13] Harriet was definitely middle-class. Henry went on to become a professor at the College of New Jersey (now Princeton University) in 1832 and as first Secretary of the Smithsonian Institution in 1846.

At this point in this presentation, I could explore Henry's scientific contributions in great detail. He was best known for his work in electromagnetism. He developed exceptionally strong electromagnets, invented the first electric motor incorporating the continual reversal of polarity of an electromagnet and he discovered that an electric current could be generated by magnetism; specifically, he identified both mutual electromagnetic induction (the action of an ordinary magnet on a conductor) and electromagnetic self-induction (the action of a current on itself). Much of this was done while he was at the Albany Academy. He examined the problem of lateral discharge, induction over long distances, and discovered the concept of the transformer. All in all, he provided the scientific foundation upon which the telegraph and the telephone, and hence the communication revolution of the nineteenth century, rested. Henry's role in the development of these inventions was not merely through publication. He advised both Samuel F. B. Morse and Alexander Graham Bell on technical issues at pivotal moments, Morse while Henry was at Princeton and Bell while Henry was at the Smithsonian.[14]

Henry also published on optics, molecular forces, and terrestrial magnetism. Using a thermoelectric device, he measured the temperature difference between a sunspot and the solar surface, the first time this difference was empirically decided. He made major contributions to the field of acoustics. And he carried out an empirical study of phosphorescence—a phenomenon now recognized as ultraviolet light emission.

In the context of the story I want to tell, however, the significance of Henry's success as an experimenter while at Albany is not his specific contributions to knowledge, but that it led to his appointment in Princeton, aided in part because he was a Presbyterian . And it was while he was at Princeton that he became part of the extended scientific community in Philadelphia and met and became friends with his partner in later transforming science in Washington, D.C., Alexander Dallas Bache .[15]

There is no question that Bache is less known than Henry, although in many ways equally important to the history of American science. His background could not have been more different than Henry's. Bache (1806-1867), born in Philadelphia, was the great-grandson of Benjamin Franklin. On his mother's side he was connected to the leadership of the Democratic Party in Pennsylvania. His grandfather, Alexander James Dallas had been Secretary of the Treasury under President Madison and his uncle George Dallas was President Polk's Vice President. He graduated first in his class at West Point in 1825 (at age nineteen). After three years in the Army, he resigned to join the faculty at the University of Pennsylvania. And it was in November 1833 that he met Henry when Henry visited Philadelphia.[16]

Bache left Philadelphia in 1843 to become superintendent of the United States Coast Survey, which, under his leadership, became the largest employer of scientists in the antebellum United States. This was the first step in the movement of the center of gravity of science in this country from Philadelphia to the nation's capital. A second step occurred when Bache was named to the first Board of Regents of the

Smithsonian Institution and successfully maneuvered to have Henry named the first secretary.

Between them, Henry and Bache dominated Washington science for a generation. But Bache was a government employee, dependent upon an appropriation for his agency, and to a certain extent, identified with the Democratic Party. He also became incapacitated in 1864 and remained so until his death in 1867. Henry, on the other hand, was free of direct dependency upon the appropriation and the federal hierarchy and had independence to be a government advisor. Although deliberately non-partisan, he had a long personal history of relationships with politicians going back to his Albany days and had friends among politicians from both the Democratic and Whig (and later, Republican) parties.[17] Perhaps naively, Henry believed that politicians could be educated as to the importance of science and would respond positively to requests by the scientific community:

> If the scientific men of the country will only be properly united they can do much for the advance of their persuits [sic] through assistance from Congress. Politicians as a class are timid except when they have an object which they know is worthy and in the advocating of which they are sure of being sustained by authority.[18]

An example of Henry's efforts to educate politicians were his interactions concerning Charles Darwin and the theory of evolution. Unfortunately, we do not know the context, but in 1865 Henry loaned Congressman Thaddeus Stevens a copy of *Origin of Species.* In his cover letter of December 13, Henry told Stevens that he considered the book "as one of prominent productions of the day—destined to have a great influence on the study of the animal and vegetable productions of the earth as well as on the tendancy [sic] of the human family so far as its members are influenced by external nature."[19] In a conversation with Senator Timothy Otis Howe in 1874, Henry once again argued for the scientific validity of Darwin's theory, but hesitated about its application to the human race, as Darwin had done in his *Descent of Man.*[20]

While Secretary of the Smithsonian Henry worked to expand existing federal programs and activities to include science and to utilize science to influence government policy. We can identify Henry's role as an advisor to the executive branch as early as Millard Fillmore's administration,[21] and this continued through subsequent administrations. Initially, Henry relished his power to influence the executive branch, once bragging that the Secretary of the Interior told him that "if the request . . . was backed by myself that would be sufficient."[22] Soon, however, he began to ration his influence or at least try to deflect requests for his assistance, informing a former student, "I am so much applied to for recommendations for positions under government that . . . were I to give one in every ten of those requested I would be bankrupt in conscience and influence.[23]

One can recognize two areas in which Henry's role as advisor truly highlighted how science could benefit the nation. The first was the Civil War. I believe the Civil War was the first in which science was part of the national arsenal, the first time the President relied upon a science advisor in wartime, and the first in which science influenced a significant U.S. policy. In what was a contradiction of Sherman's later positive view of the victories of science, Henry argued in the *Annual Report of the Smithsonian for 1862* that science could contribute to the war effort and in turn, science benefitted from war:

> Although the immediate object of war is the destruction of life and property, yet a state of modern warfare is not a condition of evil unmingled with good. Independent of the political results which may flow from it, scientific truths are frequently developed during its existence of much theoretical as well as practical importance. The art of destroying life, as well as that of preserving it, calls for the application of scientific principles, and the institution of scientific experiments on a scale of magnitude which would never be attempted in time of peace. New investigations as to the strength of materials, the laws of projectiles, the resistance of fluids, the application of electricity, light, heat, and chemical action, as well as of aerostation, are all required.[24]

I am going to highlight three examples of how science served the nation during the Civil War. But first, Let us put to rest Carl Sandberg's infamous Henry arrested as traitor story. In Sandberg's telling, without documentation, Henry had been arrested by an army officer for signaling the Confederates from one of the towers of the Smithsonian. The officer brought Henry to the White House so Lincoln could confront him, just to be told by the President that the night before he had joined Henry at the Smithsonian to observe the testing of new army signals.[25] No where in Henry's correspondence or diary entries is this event mentioned. But what does ring true is that Henry and Lincoln did interact. In fact, one of Lincoln's friends, the journalist Noah Brooks, tells a story about the lights from the Smithsonian being mistaken for traitorous signals by a Washington shopkeeper, who rushes to tell the President. The punchline here is that Henry is sitting in the White House with Lincoln when the accuser appears.[26]

My first example of Henry as advisor during the Civil War is one in which science was called upon for information, but it is also an example of the lack of certainty as to what extent the science actually influenced Presidential policy. Ambrose W. Thompson controlled a large tract of land on the Pacific side of what is now Panama near the Costa Rico border. He urged the government to use the land for the resettlement of the enslaved African-Americans to be freed under the Confiscation Act. The mining and sale of coal deposits on this land were to provide financial stability to the colony. In May of 1862 Henry was asked by Lincoln and Secretary of State William Henry Seward to see if the coal would in fact support a colony. Henry thought not (his first love had been geology, not physics, and he had lectured on it at Princeton, so he knew geology), but to be sure he turned to J. Peter Lesley, one of the country's leading geologists, for an independent opinion. Lesley confirmed Henry's evaluation that the coal was not of commercial value. Henry provided Seward with both a written and an oral report making this point and was shocked to learn subsequently that Lincoln was reluctant to give up the idea of resettlement. It was Seward, who had found great opposition to the idea among Central American nations, who persuaded Lincoln to drop the idea. International politics, not science, was probably the tipping point.[27]

A clearer example of how science, both pure and applied, aided the war effort, is the Permanent Commission of the Navy. In the early years of the war Henry had been asked by various departments of the government to provide advice on a variety of topics dealing with military technology. To ease his burden he asked the Navy to establish a commission of civilian scientists and military officers. In February 1863 Secretary of the Navy Gideon Welles established the Commission, consisting of Henry, Bache and Admiral Charles Henry Davis. They were joined later by Bache's assistant, Joseph Saxton, and General H. G. Barnard. The Commission held over one hundred meeting, evaluated more than three hundred proposals and inventions, and issued more than two hundred and fifty reports. It evaluated warship designs, signaling systems, torpedoes, underwater guns, and other ordnance. Most of these proposals were rejected as impractical or lacking improvement on established methods, although a small percentage were recommended for testing.[28]

Henry's most important contributions to the nation before, during, and after the Civil War just might have been the twenty-five years he spent on the United States Light-House Board. In the fall of 1852 President Fillmore appointed Henry as one of the six members of the newly constituted Board. Henry was chair of its committee on experiments until his death and from 1871 to 1878 he was also chair of the Board.[29]

The Light-House Board oversaw a government activity that spanned the country and had an annual appropriation in 1852 of $500,000 (in contrast the Smithsonian operating budget for that year was just under $31,000 [*Smithsonian Annual Report for 1852,* p. 51]). Henry took his responsibilities very seriously, spending anywhere from six weeks to two months annually on Light-House Board business. The Board had responsibility for the construction and repair of lighthouses, the establishment of procedures, and the procurement of equipment. Henry helped replace a politically partisan approach to applied science and technology with one based on merit, experimentation, and expertise. Under his leadership the Board became a center for some basic and a considerable amount of applied research in optics, thermodynamics, and acoustics.[30]

Henry's Civil War research centered on applied thermodynamics, specifically, lard oil. Due to Confederate raiders preying on the U.S. whaling fleet, the price of sperm oil, used in lighthouses as fuel for the lamps, sky-rocketed. Henry was asked to find a substitute and he discovered that lard oil would work. After the war he continued these experiments, looking into the possibility of substituting petroleum products or electricity. In 1864 he projected saving the government $40,000. After his death Representative Garfield credited Henry's work with saving over $100,000 a year. Asa Gray of Harvard totaled it at one million dollars.[31]

However, because this meeting is to commemorate the activities of a scientific society, I am going to focus the rest of my time on Henry's participation in scientific organizations and his perception of their importance for the advancement of science. Henry participated in three types of scientific organizations. There were the informal gatherings, scientists gathering for good food, drink, and talk. There were the organizations with a large physical footprint, such as libraries, meeting halls, and other elements which supplemented formal presentations. And there were organizations like the PSW, which existed only to host and disseminate in some manner scientific presentations.

What value were such gatherings of scientists? Henry answered this question three times, once in mid-career, twice towards the end of his life. In his 1850 presidential address to the American Association for the Advancement of Science, he told the members "that Associations of this kind have an important bearing on the advance of knowledge is proved by abundant experience. The history of discovery will show that those who have secured for themselves immortal honor by their labors in extending the boundaries of knowledge have always appeared in groups." He went on to draw an analogy from thermodynamics: "We know that the light and heat evolved from an isolated portion of fuel is far less intense than when it is burned in connection with other combustibles; each portion increases the power of the other until the whole becomes excited to an intense glow, shedding its genial influence all around. So in the reciprocal action of mind on mind there is an excitement produced highly favorable to the perception of new truths; each mind illumines the other."[32]

Two decades later, in his 1871 presidential address to the PSW, Henry returned to this theme, basing his judgment on what was a lifetime of participation in such groups: "It is mainly through the influence exerted and the assistance rendered by such associations, that science is advanced and its results given to the world. Man is a sympathetic being, and no incentive to mental exertion is more powerful than that which springs from a desire for the approbation of his fellow men; besides this, frequent interchange of ideas and appreciative encouragement are almost essential to the successful prosecution of labors requiring profound thought and continued mental exertions. Hence it is important that those engaged in similar pursuits should have opportunities for frequent meeting at stated periods."[33]

In a later address, just months before his death, Henry again reflected on the role of the organization in science. He credited these organizations with both diffusing and increasing knowledge: "It [he was speaking specifically of the PSW] tends to keep alive an active spirit of scientific advancement, not only to diffuse a knowledge of discovery among its members, but also to stimulate—by friendly criticism and cordial sympathy to new efforts in the way of explorations into the unknown."[34] The increase was dependent upon the encouragement of open, but as Henry had warned in his earlier address, constructive discussion. "Free critical discussion" of any communication presented to it was "an essential feature of a scientific society." But "critical discussion" had to have as its objective the advancement of science. Henry warned that a scientific gathering should avoid "merely verbal criticism, undue harshness on one hand and unmerited praise on the other, regard being had to truth rather than to victory or mutual admiration."[35]

Henry had a further charge to the new PSW in his 1871 presidential address. It could not simply be a place for those doing science to exchange ideas. It also had to disseminate the work of its members beyond the walls of the meeting hall. Scientific societies had to produce communications of some sort. For the nineteenth century, this meant proceedings or transactions. For the twenty-first, a posting on the internet. But diffusion was essential.[36]

Historically, every learned society, scientific or otherwise, faced three burning questions: who could become a member, who could be an officer, and who could publish in its publications. Is interest in science sufficient to join or must one demonstrate the ability to contribute to the advancement of knowledge? Are the leaders simply locally or nationally prominent individuals or are they scientists? Is publication open or subject to peer review?

There are many stories to be told. In the antebellum United States there is the ill-fated National Institute for the Promotion of Science, a local society with national pretensions, with its leadership drawn primarily from the ranks of politicians.[37] Or its successful rival, the American Association for the Advancement of Science, with a relatively open membership but leadership drawn from the ranks of the leaders of the scientific community and a growing strict control over publication.[38] There is the turn of the twentieth-century American Astronomical Society, with the debate over whether to restrict membership to professional scientists or allow amateurs to join.[39]

We are going to focus on the societies that Bache and Henry were instrumental in establishing or leading. Bache and Henry were dedicated to the upgrading of the scientific endeavor in the United States and both men used informal and formal groups in their efforts. First there was "the Club" in Philadelphia. According to Henry, the Club "has no name and must be kept secret so as to excite no jealousy." [40] It first met in December 1834 and continued meeting three or four times a year until—well, we don't know when. It is hard to specify when secret societies dissolve, but we think it survived until Bache moved to Washington. Membership was limited to the leading physical scientists—physicists, geophysicists, astronomers, meteorologists, geologists, and chemists—who lived conveniently near Philadelphia. This was an elite group, which provided a more focused audience for scientific discussion than was available at the American Philosophical Society.[41]

In Washington, Bache and Henry belonged to the Lazzaroni, Organized and led by ache, the Lazzaroni was more limited in membership but had a wider geographical distribution than the Club. Harvard professor Louis Agassiz, University of Pennsylvania professor

John Fries Frazer, and Yales' James Dwight Dana were among its members. Essentially a dining club and lobbying group which sought to encourage professionalization and specialization in US science, it also attempted to further the employment opportunities and scientific reputation of its members. The group fell apart because of Henry's refusal to go along with this latter program.[42]

The Lazzaroni did provide one gift to American and Washington science: the National Academy of Sciences. First proposed in 1858 by Louis Agassiz, the idea of an elite organization of scientists who would provide counsel to the government came to fruition in 1863 through the effort of Bache, Agassiz, and Charles Henry Davis and the cooperation of Senator Henry Wilson of Massachusetts. One very controversial provision of the legislation named the original fifty members. Another restricted the membership to fifty, so that only death or resignation could bring in fresh blood. The original fifty reflected the views of the Lazzaroni and not the true nature of the American scientific community. It was skewed towards the physical sciences and left out some of the country's leading scientists because of jealousy or hostility on the part of the members of the Lazzaroni. But one aspect of the membership reminds us what Bache and Henry (although he had not participated in the selection of the fifty) thought about the role of science. Seventeen of the members (if you include Henry) were either military officers or government employees. For a third of the members, science explicitly meant service.[43]

Hostility among those left out (and their allies), indifference on the part of the Federal government to the opportunity provided for scientific support, and indifference among the members all appeared to doom the National Academy, especially with Bache incapacitated. Ironically, Bache's death seems to have saved it. At the urging of Bache's widow, who saw the National Academy as part of Bache's legacy, Henry reluctantly accepted the presidency in 1868. And this kept it alive. In 1870, at Henry's urging, Congress removed the fifty-member limit. By 1873, thirty-three new members had been elected. By Henry's death membership had reached one hundred.[44]

Despite its Washington roots, the nineteenth-century National Academy was a national institution and the National Institute had been dissolved in 1862. What sort of organization did the local scientist have for scientific exchange?

The answer for many years was the "Saturday Club." Titian Ramsey Peale and George C. Schaeffer, two examiners at the U.S. Patent Office, were the movers behind the Saturday Club, which met between 1854 and 1871.[45] It was a gathering of scientists and those interested in science for, in Henry's words, "social meetings for the discussion of scientific subjects and for general scientific conversation."[46] The Saturday Club was, for Henry, "a source of pleasure and profit to me. Questions are frequently proposed and observations made, which give rise to interesting associations.[47]

However, in the same diary entry I have just quoted Henry complained about the limitations of the Saturday Club:

> This club consists of a few gentlemen interested in physical science, who meet in turn at each others' houses and discourse on scientific subjects in an informal way. It would be much more profitable were each member in turn obliged to present a subject in the form of a communication prepared for the purpose, but although I have repeatedly proposed this, and have endeavored to establish the custom, by example, the members are too much occupied in their several compulsory occupations to be willing to be tied up to a performance of the kind proposed.[48]

Henry continued to be dissatisfied with the Club in the post-bellum years. Washington had become the center of science in this country and it needed some sort of formal scientific organization. He toyed with the idea of having the local scientific community linked with the National Academy, noting in an 1868 diary entry that "Washing[ton] contains a larger number of men connected with scientific operations than any other city all of which should be affiliated with the Academy."[49] In a March 22, 1870 letter to Asa Gray of Harvard, Henry thought out loud about the possibility of affiliation with the National Academy: "We are much in want of a local society for uniting all the men of science in this city in one organization, and this may perhaps be done through the National Academy, although the manner of accomplishing the project is not quite clear to my mind."[50] Henry went so far as to have a meeting on the subject the following month, documented in his diary entry of April 11, in which the idea of "the formation of a scientific society in the city of Washington as an auxiliary to the academy" was considered and rejected in favor of electing local "associates" to the Academy "as such a sufficient number to hold meetings in Washington."[51]

The actual solution to a lack of a local scientific organization was quite different. On February 7, 1871, a petition signed by forty-three men, primarily scientists and engineers, but including General William T. Sherman and Chief Justice Salmon P. Chase, was presented to Henry, requesting him to "preside at a meeting which they propose to hold for the purpose of forming a society, having for it object, the free exchange of views and studies on scientific subjects, and the promotion of scientific inquiry among its members.[52] In response, Henry called a meeting "to form a scientific Society for the city of Washington."[53]

For some reason, this meeting seems to have disappeared from the official history of the PSW. The above quotation is taken from Henry's diary entry of March 6. There is no question this meeting was about the formation of the PSW, although Henry calls it the "Washington Philosophical Society." Henry even tells us what the leadership would consist of: "a President four vice Presidents two secretaries and an executive committee of nine members," which is how the PSW was organized. But the *Bulletin of the Philosophical Society of Washington* states that March 13 was the first meeting.[54] So is this the 2501st meeting?

The important takeaway from this is that when the community wanted to create a local scientific society in Washington, it went to Henry. In response, Henry took leadership of the society. He chaired both the March 6 and 13 meetings, was elected the first president of the PSW, and was reelected every year until his death. The PSW held a special place in his life. It was at the November 24, 1877, meeting that Henry gave his last scientific presentation. Less than two weeks later, on December 5, he was struck by an attack of paralysis which marked the beginning of his final illness.

In this last presentation[55] Henry prefaced his remarks with praise of the PSW. The PSW "is intimately connected with the intellectual development of the City of Washington. . . It tends to keep alive an active spirit of scientific advancement, not only to diffuse a knowledge of the progress of discovery among its members, but also to stimulate by friendly criticism and cordial sympathy to new efforts in the way of explorations into the unknown."[56] He then expounded his theory of what are the requisites of a good scientist and the proper manner of research. His scientific exposition was presented as an illustration of good research to illuminate the points he had made earlier. It incorporated observations about abnormal phenomena of sound which he had made while investigating fog-signals for the Light-House Board. To the end of his life, Henry thought of science as useful and in service to the nation.

Now a postscript. Seven months after Henry's death, John Wesley Powell, explorer, geologist and anthropologist, called a meeting at his home to launch a social club for scientist. The ultimate result was the organization whose clubhouse we are meeting in. The Cosmos Club was not a direct rival of the PSW. Rather, it complimented it. And to forestall any bad feelings, the membership of the PSW was invited to become charter members of the Cosmos Club.[57]

My own speculative belief is that Powell recognized there was a void left when Henry died. Who would take on his mantle as the new leader of the scientific community in the Halls of Congress and in the offices of Cabinet members? I think Powell thought he could and saw the Cosmos Club as part of his plan. But that would be for another talk.

Almost 153 years after those forty-three men petitioned Henry, the PSW continues, through its presentations, to carry out Henry's charge to the organization: to diffuse knowledge and stimulate scientific advancement. Here's to another 2500 presentations!

[1] Lin-Manuel Miranda and Jeremy McCarter, *Hamilton: The Revolution* (New York, 2016), p. 16

[2] I have previously explored some of the themes in this presentation in a previous lecture to the Philosophical Society of Washington in 1997: "'Frequent Interchange of Ideas and Appreciative Encouragement': The Role of Scientific Societies in the Life of Joseph Henry." Also see Marc Rothenberg, "Learned Societies and the Birth of American Science," in Lila Vekerdy, *The Era of Experiments and the Age of Wonder: Scientific Expansion from the Seventeenth to the Nineteenth Centuries* (Washington, D.C., 2015, pp.23-30.

[3] *A Memorial of Joseph Henry* (Washington, D.C., 1880), p. 12.

[4]*Bulletin of the Philosophical Society of Washington, 2:*323.

[5] *Memorial, p. 97.*

[6] *Memorial,* p. 118.

[7] *Memorial,* p. 119.

[8] Unless otherwise noted, information on Henry's life and work will be taken from Albert E. Moyer, *Joseph Henry: The Rise of an American Scientist* (Washington, D.C., 1997).

[9] Nathan Reingold, ed., *The Papers of Joseph Henry* (Washington, D.C. 1972), *1*:3. Hereafter *PJH.*

[10] Henry's copy of the book survives in the Alexander Graham Bell-Joseph Henry Library in the Dibner Library of the History of Science and Technology, Smithsonian Institution.

[11] P. 1.

[12] P. 2.

[13] Marc Rothenberg, ed., *PJH* (2004), *10:*548-549.

[14] For specific instances of Henry's consultations with Morse and Bell, see Rothenberg, ed., *PJH* (1992), *6:*9-10, 21-22 (Morse) and Rothenberg, ed., *PJH* (2007), *11:*523-529 (Bell).

[15] For Bache I have drawn upon material in *PJH,* the biography by Axel Jansen, *Alexander Dallas Bache: Building the American Nation through Science and Education in the Nineteenth Century* (Frankfurt, 2011), and Hugh Richard Slotten, *Patronage, Practice, and the Culture of American Science: Alexander Dallas Bache and the U.S. Coast Survey* (Cambridge, Eng., 1994).

[16] Reingold, ed., *PJH* (1975), *2:*88, 108.

[17] I have explored Henry's role as science advisor to the Federal government in two unpublished presentations. "Science Advisor to the Federal Government: A Nineteenth-Century Experiment," Washington Seminar on American History and Culture, January 1998; and "Science and Union" Cosmos Club Civil War Group, August 2017.

[18] Henry to Asa Gray, February 21, 1849, Rothenberg, ed., *PJH* (1996), *7:*480.

[19] Rothenberg, ed., *PJH,* (2004), *10:*566.

[20] Howe to Grace T. Howe, January 3,1874, Howe Manuscripts, State Historical Society of Wisconsin.

[21] Rothenberg, ed., *PJH* (1998), *8:*257.

[22] Henry to Alexander Dallas Bache, October 25,1851, Bache Papers Smithsonian Institution Archives.

[23] Henry to John Miller, August 27, 1852, Charles Roberts Autograph Letters Collection, Haverford College Library.
[24] P. 13.
[25] Sandburg, *Abraham Lincoln: The Prairie Years and the War Years,* rev. ed., 3 vols. (New York, 1959, *2:*113-114.
[26]Noah Brooks, *Washington in Lincoln's Time* (New York, 1895), pp. 11-13.
[27] Rothenberg, ed., *PJH, 10:*268-269, 279.
[28] Nathan Reingold, "Science in the Civil War: The Permanent Commission of the Navy Department," *Science, American Style* (New Brunswick, 1991), pp. 79-95; Rothenberg, ed. PJH, 10:309-313.
[29] *Annual Report of the Light-House Board for the Year 1882,* pp. 130-131.
[30] This discussion is based on a reading of the annual reports of the Light-House Board.
[31] Rothenberg, ed., *PJH, 10:*xxxiii, 375; *11:*xli-xlii; *Memorial,* pp. 69-70, 96-97.
[32] Arthur P. Molella, et al., *A Scientist in American Life: Essays and Lectures of Joseph Henry* (Washington, D.C., 1980), 38-39.
[33] "The Organization of a Scientific Society," *Bulletin of the Philosophical Society of Washington, 1:*6.
[34] "The Methods of Scientific Investigation and Its Application to Some Abnormal Phenomena of Sound," *Bulletin of the Philosophical Society of Washington, 2:*162.
[35] "The Organization of a Scientific Society," *Bulletin of the Philosophical Society of Washington, 1:*9.
[36] *Ibid.,1:*10.
[37] Sally Gregory Kohlstadt, "A Step Towards Scientific Self-Identity in the United States: The Failure of the National Institute, 1844," *Isis*, 62 (1971): 339-362.
[38] Kohlstedt, "Creating a Forum for Science: AAAS in the Nineteenth Century," in Kohlstedt, Michael M. Sokal, and Bruce V. Lewenstein, *The Establishment of Science in America: 150 Years of the American Association for the Advancement of Science* (New Brunswick, 1999), pp. 7-49.
[39] Marc Rothenberg, "Organization and Control: Professionals and Amateurs in American Astronomy, 1899-1918," *Social Studies of Science,* 11 (1981):305-325; Marc Rothenberg and Thomas R. Williams, "Amateurs and the Society during the Formative Years." In David H. DeVorkin, ed., *The American Astronomical Society's First Century* (Washington, D.C., 1999), pp. 40–57.
[40] Henry to John Torrey, December 20, 1834, Reingold, ed., *PJH, 2:305.*
[41] *Ibid., 2:*290-291.
[42] At one time the Lazzaroni drew considerable attention from historians. See, e.g., Mark Beach, "Was There a Lazzaroni?" In George H. Daniels, editor, *Nineteenth-Century American Science: A Reappraisal* (Evanston, 1972), pp. 115–132; Robert V. Bruce, *The Launching of Modern American Science, 1846–1876* (New York, 1987), pp. 220-225; Lillian Miller, et al., *The Lazzaroni: Science and Scientists in Mid-nineteenth Century America* (Washington, D.C., 1972).
[43] Bruce, pp. 302-304; Frederick W. True, ed., *A History of the First Half-Century of the National Academy of Sciences, 1863-1913* (Washington, D.C., 1913).
[44] Bruce, pp. 315-317; Rothenberg, ed., *PJH, 11:172.*
[45] Rothenberg, ed., *PJH, 9:*167.
[46] Henry Locked Book, December 23, 1854, *Ibid., 9:*170.

[47] Henry Locked Book, March 8, 1862, Rothenberg, ed., *PJH, 10:*237.
[48] *Ibid., 10:*236-237.
[49] January 25, 1868, *Ibid., 11:*172
[50] *Ibid., 11:*284-285.
[51] *Ibid., 11:*286-287.
[52] The manuscript petition is in Philosophical Society of Washington, RU 7079, Smithsonian Institution Archives. It was printed in the *Bulletin of the Philosophical Society of Washington, 1:*19-20.
[53] Henry Desk Diary, March 6, 1871, Rothenberg, ed., *PJH, 11:*345.
[54] *1:*20.
[55] "Methods of Scientific Investigation and Its Application to Some Abnormal Phenomena of Sound," *Bulletin of the Philosophical Society of Washington, 2:*162-174.
[56] *Ibid.,* p. 162.
[57] Wilcomb E. Washburn, *The Cosmos Club of Washington* (Washington, D.C., 1995), pp. 15-20.

Invitatory Letter
Requesting Joseph Henry to Form a Scientific Society
February 7, 1871

"Invitatory Letter." 1962. *Bulletin of the Philosophical Society of Washington* (16): 156.

Prof Joseph Henry LL.D.

The undersigned respectfully request you to preside at a meeting which they propose to hold for the purpose of forming a society, having for its object, the free exchange of views and studies on scientific subjects, and the promotion of scientific inquiry among its members.

7 February 1871

M C Meigs
Benjamin Peirce
Theo. Gill.
Peter Parker
F. B. Meek
T R Peale.
Wm B. Taylor
Chas. A. Schott
E. B. Elliott
F. V. Hayden
Albert J. Myer
A A Humphreys
Thos Lincoln Casey
Jno: G. Parke

J. E. Hilgard
J. H. Lane
S F Baird
Horace Capron
Thomas Antisell
J. J. Woodward
J S Billings
J K Barnes
C. H. Crane
George A. Otis
B F Craig
Walter L. Nicholson
B. F. Sands

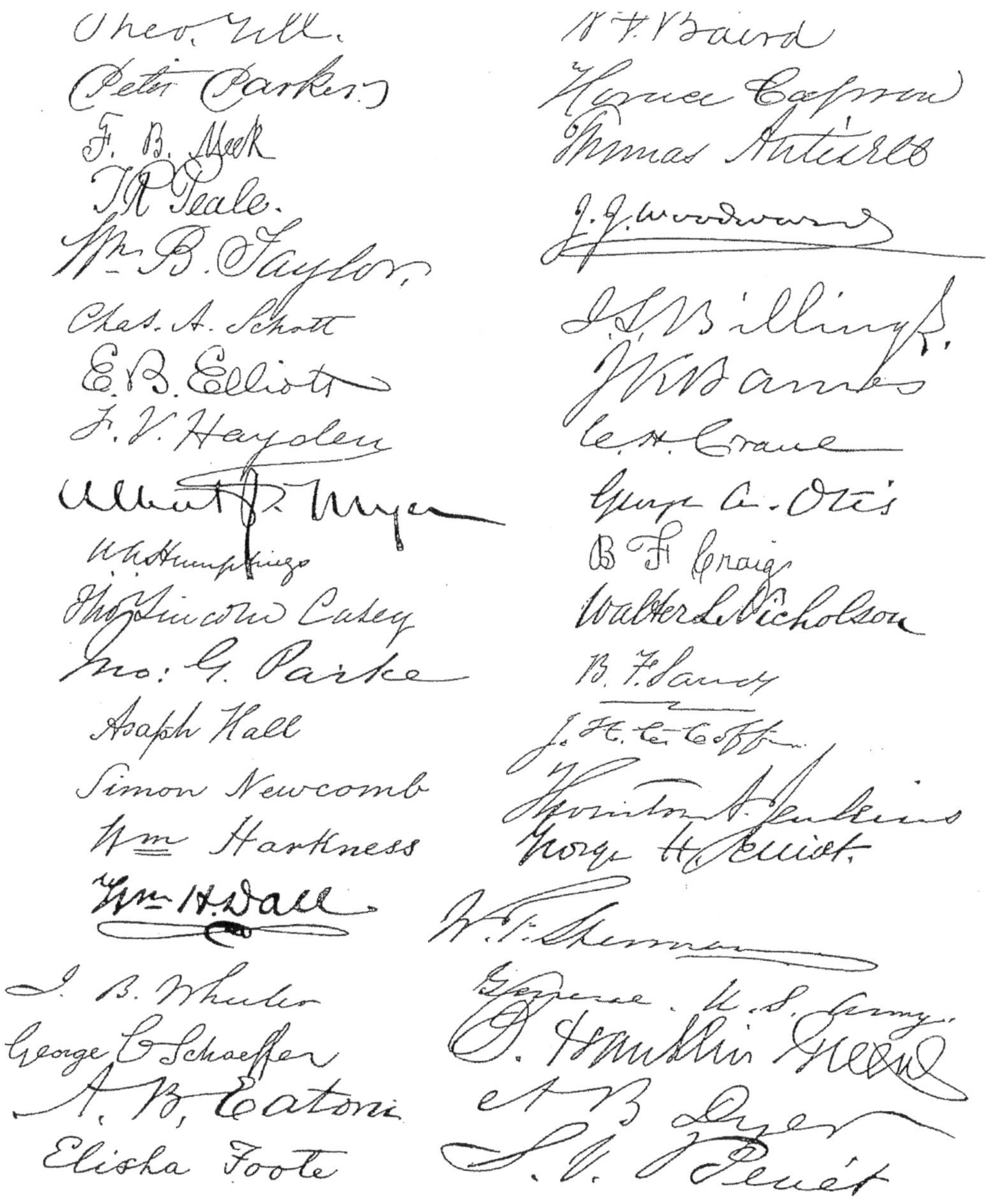

DRAFT OF THE INITIATORY LETTER DATED FEBRUARY 7, 1871

The above document appears to be an earlier draft of the Initiatory Letter which was at the origin of the foundation of The Philosophical Society of Washington by 44 members of the Saturday (or Scientific) Club on March 13, 1871. Most probably some of the members of the Saturday Club signed this draft after February 7 (which was a Tuesday). The names of Elisha Foote, Wm. Harkness, and Simon Newcomb seem to have been written in by another person (possibly Asaph Hall). The original document shows that Wm. H. Dall and George A. Otis placed their signatures over their names written in by another person. Salmon P. Chase, whose name appears on the final document, did not sign this draft and W. T. Sherman's signature is followed by the words: General U.S. Army.

The text of the above draft includes the words "and studies" which are omitted in the final document.

INITIATORY LETTER

The Philosophical Society of Washington had its origin in the following initiatory letter:

Prof. Joseph Henry, LL.D.

The undersigned respectfully request you to preside at a meeting which they propose to hold for the purpose of forming a society, having for its object the free exchange of views on scientific subjects, and the promotion of scientific inquiry among its members.

M. C. Meigs,	Horace Capron,
Benjamin Peirce,	Thomas Antisell,
Theo. Gill,	J. J. Woodward,
Peter Parker,	J. S. Billings,
F. B. Meek,	J. K. Barnes,
T. R. Peale,	C. H. Crane,
Wm. B. Taylor	George A. Otis,
Chas. A. Schott,	Albert J. Myer,
E. B. Elliott,	A. A. Humphreys,
F. V. Hayden,	Asaph Hall,
J. E. Hilgard,	Simon Newcomb,
J. H. Lane,	Wm. Harkness,
S. F. Baird,	B. F. Craig,
Walter L. Nicholson,	J. H. C. Coffin,
Wm. H. Dall,	Thornton A. Jenkins,
B. Franklin Greene,	George H. Elliot,
S. V. Benét,	W. T. Sherman,
George C. Schaeffer,	A. B. Dyer,
Thos. Lincoln Casey,	J. B. Wheeler,
Jno. G. Parke,	A. B. Eaton,
B. F. Sands,	Elisha Foote,

Salmon P. Chase.

The original of the above letter has not yet been located, however, the letter is quoted in earlier documents (including Vol. 1 of the Bulletin) *and was apparently presented to Joseph Henry on March 12, 1871.*

The document reproduced on the fold-out insert was found recently in the archives of the Society and appears to be a precursor of the Initiatory Letter.

156

Articles of Incorporation
District of Columbia
Philosophical Society of Washington
Filed May 20, 1901

Articles of Incorporation.

was organized in the city of Washington, District of Columbia, for scientific purposes, March 13, 1871, as an unincorporated association and has so continued to the present time, and is now desirous of becoming incorporated; now, therefore

We, the undersigned, founders, and members of said unincorporated association, and pursuant to instructions therefrom, being citizens of the United States and a majority of whom are citizens of the District of Columbia do hereby associate ourselves as a body corporate, for the above mentioned purposes, under the general incorporation acts of the Congress of the United States enacted for the District of Columbia; and we do hereby certify, in pursuance of said acts as follows:-

First. The name or title, by which such society shall be known in law is the Philosophical Society of Washington.

Second. The term for which said society is organized is nine hundred and ninety nine years.

Third. The particular business and objects of the society are the promotion of science, the advancement of learning, and the free exchange of views among its members on scientific subjects.

Fourth. The number of its managers for the first year of its existence shall be twenty-five.

[In

In testimony whereof we have hereto set our names and affixed our seals, at the city of Washington, in the District of Columbia on the fifteenth day of May, 1901.

Wm H Dall founder (Seal)
Theo. Gill founder (Seal)
Simon Newcomb founder (Seal)

Cyrus Adler (Seal)
Marcus Baker (Seal)
Louis A. Bauer (Seal)
Frank H. Bigelow (Seal)
F. W. Clarke (Seal)
Wm. A. DeCaindry (Seal)
Robert Fletcher (Seal)
G. K. Gilbert (Seal)
James H. Gore (Seal)
John G. Hagen (Seal)
John F. Hayford (Seal)
George W. Littlehales (Seal)

Charles F. Marvin (Seal)
W. W. Paul (Seal)
J. W. Powell (Seal)
Richard Rathbun (Seal)
Geo. M. Sternberg (Seal)
Otto H. Tittmann (Seal)
F. W. True (Seal)
Chas D. Walcott (Seal)
(Seal)
Charles K. Wead (Seal)
Isaac Winston (Seal)

District of Columbia ss.

I, Henry E. Cooper, a Notary Public in and for the District of Columbia do hereby certify that Wm. H. Dall, Theo. Gill, Simon Newcomb, Cyrus Adler, Marcus Baker, Louis A. Bauer, Frank H. Bigelow, F. W. Clarke, Wm. A. DeCaindry,

Robert Fletcher, G. K. Gilbert, James H. Gore, John G. Hagen, John F. Hayford, George W. Littlehales, Charles F. Marvin, H. M. Paul, J. W. Powell, Richard Rathbun, Geo. M. Sternberg, Otto H. Tittmann, F. W. True, Chas. D. Walcott, J. E. Watkins, Charles K. Wead, and Isaac Winston, parties to the foregoing Articles of Incorporation, dated the fifteenth day of May, A.D., 1901, and to which their signatures are affixed, are to me personally known; that they personally appeared before me in the said District of Columbia, and acknowledged to me that they had executed the said Articles of Incorporation for the uses and purposes therein set forth.

Given under my hand and official seal this eighteenth day of May, A.D. 1901.

Henry E. Cooper
Notary Public

RECORDER OF DEEDS
DIST. OF COL.

MAY 20 1901

Philosophical Society

Incorporation

RECEIVED [illegible]
MAY 20 A.D. 1901
2.56 P.M. and rec[illegible]
Liber No. 9 folio 277 [illegible]
one of the ~~Land~~ Records of
the District of Columbia.
[illegible signature]
Recorder.

$2.00

1st Meeting of the Philosophical Society of Washington
March 13, 1871
Regents Room, Smithsonian Institution
Professor Joseph Henry in the Chair

1st Meeting. 1874. *Bulletin of the Philosophical Society of Washington* (1): 20.

1st Meeting. March 13, 1871.

Prof. Joseph Henry in the Chair.

In response to this call, a meeting of the subscribers thereto was convened and held at the Smithsonian Institution, in the Regent's room, on Monday, March 13, 1871. The outline of a Constitution was adopted, and under it the following gentlemen, who collectively should constitute a General Committee for the transaction of the business of the Society, were elected officers:—

PRESIDENT.

Joseph Henry.

VICE-PRESIDENTS.

M. C. Meigs, J. E. Hilgard, Horace Capron, Wm. B. Taylor.

TREASURER.

Peter Parker.

SECRETARIES.

B. F. Craig, Theodore Gill.

MEMBERS AT LARGE OF THE GENERAL COMMITTEE.

Thomas Antisell, J. H. C. Coffin, S. Newcomb, S. F. Baird, E. B. Elliott, W. T. Sherman, T. L. Casey, T. A. Jenkins, J. J. Woodward.

The Constitution was then referred to the General Committee for verbal expression; and the Committee was also empowered to propose Rules and By-Laws for the government of the Society.

Anniversary Address of Joseph Henry
1st President of the Philosophical Society of Washington
15th Meeting
November 18, 1871

Henry, Joseph. 1874. "Anniversary Address of the President of the Philosophical Society of Washington." *Bulletin of the Philosophical Society of Washington* (1): 5-14.

The Name of the Society. "The term "Philosophical" was chosen not to denote, as it generally does, in the present day, the unbounded field of speculative thought, which embraces the possible as well as the actual of existence, but to be used in its restricted sense to indicate those branches of knowledge that relate to the positive facts and laws of the physical and moral universe." (Henry 1874, 6)

Candid and Principled Discourse is Essential. "It is an essential feature of a scientific society that every communication presented to it should be subject to free critical discussion. Such discussion not only enlivens the proceedings, but is, generally, instructive, frequently eliciting facts which, though insignificant when isolated, when brought together mutually illustrate each other, and lead, ultimately, to important conclusions. . . . Among the things to be avoided are, merely verbal criticism, undue harshness on the one hand, and unmerited praise on the other, regard being had to truth rather than to victory or mutual adulation. There is nothing, perhaps, which marks more distinctly one of the characteristics of a true scientist than the manner in which he receives and appropriates to his use the critical remarks that may be made upon his communications." (Henry 1874, 9)

Disseminate the Knowledge. "Another important matter in regard to such a society is the publication of its proceedings. If its object were merely the intellectual and moral improvement of its members, it might dispense entirely with any publication whatever – even with the announcement of its existence. If, however, it aspires to the more important office of *advancing* science, or of enlarging the bounds of thought and assisting to diffuse a knowledge of new truths, it should then publish . . . its proceedings." (Henry 1874, 10)

ANNIVERSARY ADDRESS

OF THE

PRESIDENT OF THE PHILOSOPHICAL SOCIETY OF WASHINGTON,

PROF. JOSEPH HENRY.

(Delivered November 18, 1871.)

GENTLEMEN:

I have been requested to make some remarks on the character and object of this Society which may serve to introduce it to the world through the pages of a Bulletin of its proceedings or the public journals of the day, and in compliance with this request I beg leave to submit the following.

This Society was formed by the call for a meeting of a number of gentlemen impressed with the importance of an association of a strictly scientific character in the city of Washington. At the meeting which resulted from this call a name and a constitution were adopted for the Society, and without delay, in a series of subsequent meetings, the objects of the association were prosecuted with such marked success as to fully realize the anticipations which had been entertained with regard to the enterprise. This is manifest from the number, character, and variety of the communications presented and discussed.

In regard to the name which has been chosen, "THE PHILOSOPHICAL SOCIETY OF WASHINGTON," it is proper to remark that it was adopted not without considerable deliberation. The term "Philosophical" was chosen not to denote, as it generally does, in the present day, the unbounded field of speculative thought, which embraces the possible as well as the actual of existence, but to be used in its restricted sense to indicate those branches of knowledge that relate to the positive facts and laws of the physical and moral universe. The second term, "Washington," was selected to denote the fact that the Society is a *local* establishment; that it arrogates to itself nothing on account of its position at the national capital; makes no claim to any connection with the government, nor to being, in any respect, a special representative of the science of the country.

The importance of such a society must be evident to all who are acquainted with the history of science. It is mainly through the influence exerted and the assistance rendered by such associations that science is advanced and its results given to the world. Man is a sympathetic being, and no incentive to mental exertion is more powerful than that which springs from a desire for the approbation of his fellow men: besides this, frequent interchange of ideas and appreciative encouragement are almost essential to the successful prosecution of labors requiring profound thought and continued mental exertion. Hence, it is important that those engaged in similar pursuits should have opportunities for frequent meetings at stated periods; this is more particularly the case with the cultivators of abstract science, who find but comparatively few fully capable of appreciating the value of their labors, even in a community how much soever enlightened it may be on general subjects. The students of history, of literature, of politics, and of art find everywhere men who can enter in some degree into their pursuits, and who can appreciate their merits and derive pleasure from their writings or conversation; while the mathematician, the astronomer, the physicist, the chemist, the biologist, and the student of descriptive natural history meet with few, comparatively, who can sympathize with them in their pursuits, or who have a sufficient knowledge of their particular subjects to be able to award them that intelligent appreciation and encouragement essential to their sustained and laborious efforts. To them, the world consists of a few individuals, to whom they are to look for that critical judgment of their merits which is to be finally adopted by the general public, and with these it is of the first importance that they should have more frequent intercourse than that which arises from casual meetings.

Furthermore, a society of this kind becomes a means of instruction to all its members, the knowledge of each becoming, as it were, the knowledge of the whole. Again, there is a common bond of union between all branches of science, since they all relate to the existence and laws of the same universe in which the more we extend our knowledge the more we find of "unity in the midst of infinite diversity." This connection is obvious in the relations of astronomy, mathematics, and physics, as well as in those of geology, chemistry, and biology, which are so closely related in many cases as to be separable only by conventional limits. In a society, therefore, like the one in question, embracing in its objects, as it does, all branches of science, each investigator may find others cultivating fields separated from his own by insensible degrees, from whom he can have not only full sympathy and adequate appreciation, but also, in many instances, important suggestions and essential aid.

The governing body of such a society, in order that the

organization may produce the desired effect, must be largely composed of men who, by education and experience in the processes of investigation, are justly entitled to the appellation of "scientific," and who, from their positive contributions to the science of the day, are acknowledged by the scientific world as worthy of this distinction. It is true, useful societies are formed for the self-improvement of their members by the production of essays on various subjects, or by cultivation of branches of natural history requiring no previous special training: the city of Washington, however, needs something of a higher order, namely, a society for the *advancement* of science, since in no other city in the Union are there so many men, in proportion to the population, connected with scientific pursuits, or so many facilities for scientific investigation.

The Philosophical Society of Washington, though of a local and unostentatious character, may, if true to itself and its mission, accomplish much towards increasing the reputation of the country and influencing public opinion with regard to questions of a scientific character. However wide the diffusion of general knowledge, public opinion in regard to scientific questions must eventually be determined by the authority of societies, journals, and individuals, of established scientific reputation. It is therefore of the first importance that the operations of this Society be conducted with great care, and that nothing be given to the world under its sanction which is not based upon thorough investigation or established scientific principles. We should be warned by the fate of a society established in this city some thirty years ago, which, although it included among its members a few men of true science, was under the control principally of amateurs and politicians, and therefore was unfit to discharge the duty which it claimed as one of its functions, to decide questions of a strictly scientific character. It should have been borne in mind by this association that votes on questions in science should be *weighed*, not counted! Had the proposition of the motion of the earth been decided in the days of Galileo by the popular voice, this philosopher and his friends would have been vastly in the minority. The society to which I allude, after achieving an unenviable notoriety, by assuming to be the arbiter of the science of the country, gradually sunk into oblivion, from which its memory should not be recalled except as a warning to those who would adventure in the same line.

It is an essential feature of a scientific society that every communication presented to it should be subject to free critical discussion. Such discussion not only enlivens the proceedings, but is, generally, instructive, frequently eliciting facts which, though insignificant when isolated, when brought together mutually illustrate each other, and lead, ultimately, to important conclusions. The extent to which discussions may be allowed evidently depends on the candor and temper of those who engage in them. Among the things to be avoided are, merely verbal criticism, undue harshness on the one hand, and unmerited praise on the other, regard being had to truth rather than to victory or mutual adulation. There is nothing, perhaps, which marks more distinctly one of the characteristics of a true scientist than the manner in which he receives and appropriates to his use the critical remarks that may be made upon his communications. He can, in many cases, at least, derive from them the indication that he has failed to present on some points a clear statement of his investigations;

or that in some other points his conclusions are not fully sustained by the premises. Unfortunately, it frequently happens that persons of a sensitive disposition are apt to consider criticism of the kind we have mentioned as personal attacks, and that it is as offensive to doubt the accuracy of their experiments or conclusions as it is to doubt their word. It should, however, be recollected that the most gifted are liable to err, and that these criticisms are *prior* to publication, and, therefore, of value to the permanent reputation of both the individual and the society.

Another important matter in regard to such a society is the publication of its proceedings. If its object were merely the intellectual and moral improvement of its members, it might dispense entirely with any publication whatever—even with the announcement of its existence. If, however, it aspires to the more important office of *advancing* science, or of enlarging the bounds of thought and assisting to diffuse a knowledge of new truths, it should then publish, if not quarto volumes of transactions, at least a bulletin of its proceedings. This publication should present an exposition of the organization of the society, its constitution and by-laws, give a list of the members, a synopsis of the contents of all communications submitted for consideration, and an account of important facts which may be elicited during discussions or recalled to memory at the moment by association of ideas.

Such a bulletin will enable the members of the society to publish without delay, through a proper channel, a synopsis of their investigations, and, also, minor facts and inferences not considered, in themselves, of sufficient importance to form a communication to a scientific journal or to occupy a place in philosophical transactions. Such facts are, nevertheless, frequently found to be valuable contributions to the general stock of knowledge. Were it possessed of the requisite funds, the society might establish a higher reputation by the publication of independent transactions. Inasmuch, however, as this is not the case, the next best plan should be adopted, namely, that of publishing papers in full through other channels, such, for instance, as the Smithsonian Institution, the reports of government bureaus, and scientific journals. In such cases, the bulletin should contain references as to where the articles in full are to appear, and in this respect it would do good service in assisting to make more generally known the valuable contributions to science which are diffused through voluminous executive and congressional documents not readily accessible to the scientific world.

The editing of the bulletin should be under the direction of the secretaries and a committee appointed for the purpose, and a number should be issued as often as material of the proper character and of sufficient quantity is accumulated. It should be distributed to the principal learned societies of this and other countries, and may also be presented to leading journals in this and other cities. Without at least such a publication, the society cannot have a recognized existence.

I have stated that there is no city in the United States, in proportion to the number of its inhabitants, where there are so many men of education actively engaged in pursuits connected with science as in Washington. In illustration of this remark I may refer to those who are engaged in the Coast Survey, the Office of Weights and Measures, the National Observatory, the Nautical Almanac, Patent Office, Engineer Department, Hydrographic Office, Ordnance Department, Medical Departments of the Army and Navy, Lighthouse Board, Signal Corps, Agricultural Department, Bureau of Statistics, Census Office, Bureaus of Navigation and Steam Engineering, the Smithsonian Institution, etc. etc. In addition to this, no city in the Union possesses more ample facilities, in the way of books and implements, for the prosecution of scientific research. The library of Congress, enriched by the Smithsonian Deposit with the transactions of all the principal learned societies of the world, is almost unrivalled in scientific works. If to this extensive collection we add the special libraries of the Patent Office, the Agricultural Department, the Coast Survey, the National Observatory, and of the Surgeon-General's Office, we have a collection of modern books on science, accessible to the members of the society, scarcely surpassed by the collections of the most favored cities of the old world. Nor are the articles of apparatus necessary for any line of investigation beyond the reach of any member of the Society who may possess the knowledge and skill requisite to their proper use. There is great liberality on the part of the heads of departments in the way of furnishing apparatus that may in any degree facilitate the special investigations under their direction.

Among those connected with the various organizations just mentioned, a considerable number are engaged in *original* investigations, the results of which are of interest to the scientific world, and which will be facilitated and improved by the discussions of this Society. Furthermore, in the daily operations of the different establishments, facts of scientific importance are continually becoming evident which would be lost if not preserved in the records of the Society. It is not, however, alone to facilitate operations now going on, or to preserve facts that may have been casually discovered, but, also, to suggest new investigations and to encourage others to enter the field of research who have not yet essayed their hand in this direction. In the great domain of science, there is abundant room for an infinite number of laborers of different grades of attainment and original powers of mind. A series of careful observations made with proper instruments, with regularity and precision, which requires little more than the exercise of the senses and a conscientious regard for truth, is frequently a valuable contribution to science. A series of analyses in which prescribed formulas are observed, and in the application of which no more talent is required than that which is possessed by the majority of persons of ordinary ability and education, may give results of scientific value. For the production of results of the kind mentioned, and those which are effected by the scientist who is capable of detecting hitherto undiscovered facts and developing new laws, there is room for all grades of talent and of powers of original investigation. It is astonishing how much may be done by the association of minds determined on a common pursuit; how much, under such conditions as exist in the city of Washington, may be effected in the way of directing attention to special lines of investigation, in suggesting questions to be asked of nature, and in pointing out the ready means by which the answers may be elicited, by arousing into activity talents which, without such stimulus and suggestion, would ever remain dormant.

The bane of many societies is the time consumed in details of business and in the discussion of non-essential points relative to their government. Happily, the organization adopted by this Society obviates this evil and secures the devotion of almost every evening exclusively to its legitimate purposes. For the government of men whose object is the advance of *truth*, but few rules are necessary, and these, unlike the laws of the Medes and Persians—expressed in inexorable codes—must consist of simple principles, readily adaptable to all contingencies.

In conclusion, I would say that with so many facilities as exist in the city of Washington for the pursuit of science, this Society would be derelict of duty did it fail to materially aid, through communion of thought and concert of action, the advancement of the great cause of human improvement. I am happy, however, in cherishing the opinion that the success of "The Philosophical Society of Washington" is scarcely any longer problematical, and in this I am sustained by the record of its transactions.

Annual Address of Joseph Henry
President of the Philosophical Society of Washington

131st Meeting
November 24, 1877

131st Meeting. 1880. *Bulletin of the Philosophical Society of Washington* (2): 162-175.

Membership Requires Only a High Appreciation of Science.
"While but comparatively few qualifications are necessary for admittance, yet no person is elected who is not supposed to have at least a high appreciation of science; has some familiarity with its principles, and is capable of doing something in the way of promoting the objects of the Association." (Henry 1880, 162)

131st Meeting. November 24, 1877.

The President, Joseph Henry, in the Chair.

Thirty-six members and visitors present.

The President read his annual address, as follows:—

Gentlemen, Members of the
Philosophical Society of Washington.

I beg leave to tender you my sincere thanks for the honor you have conferred upon me, and the good feeling you have manifested towards me, by my re-election as President of this Society. I say the good feeling which you have manifested towards me, because I know that there are many of your members who can much more efficiently discharge the duties of the office than I can. I may, perhaps, be allowed to say, without the charge of undue egotism, that I have never occupied any position for which I have been voluntarily a candidate. The several offices of honor and responsibility which I now hold, no less than nine in number, have all been pressed upon me without solicitation on my part, and I now begin to feel that, in view of that peculiarity of human nature so admirably exhibited in the character of the Archbishop of Granada, that I ought to diminish the number of my responsibilities, gradually leaving to others the honor and the toil of office. It is, therefore, with no feigned hesitation that I again accept the re-election to the position to which your kindness has called me.

I have, however, taken from the first a deep interest in the Society, knowing that it is intimately connected with the intellectual development of the City of Washington, and that it has a reflex influence upon every part of the United States. It tends to keep alive an active spirit of scientific advancement, not only to diffuse a knowledge of the progress of discovery among its members, but also to stimulate by friendly criticism and cordial sympathy to new efforts in the way of explorations into the unknown.

While but comparatively few qualifications are necessary for admittance, yet no person is elected who is not supposed to have at least a high appreciation of science; has some familiarity with its principles, and is capable of doing something in the way of promoting the objects of the Association.

The general mental qualification necessary for scientific advancement is that which is usually denominated "common sense," though, added to this, imagination, invention, and trained logic, either of common language or of mathematics, are important adjuncts. Nor are objects of scientific culture difficult of attainment. It has been truly said that the "seeds of great discoveries are constantly floating around us, but that they only take root and germinate in minds well prepared to receive them."

The preparation, however, is not difficult, and many possess the requisites in an eminent degree who are not aware of the fact. Genius itself has been defined as a mind of general powers determined, enthusiastically it may be, on one pursuit.

The method of discovery or scientific observation is not difficult. There is a story in a work, entitled "Evenings at Home," which produced an indelible impression on my mind. It is entitled "Eyes and no Eyes," and related to two boys who started on a walk during a warm summer afternoon. On their return, one was fatigued, dissatisfied, having seen nothing, encountered only dust and heat, while the other was charmed with his walk, which had been over the same ground, and gave a glowing account of the objects which he had met with, and of the reflections which were awakened by them. On this story, De la Bêche has founded a work, entitled "How to Observe in Geology," which I would commend to the attention of every member of this Society, while I suggest that good service would be done to the advance of knowledge were a similar work published relative to all branches of science.

The first requisite for an observer is, that his mind should be actively awakened to the phenomena of nature with which he is surrounded. Thousands of persons of excellent mental capacity pass through the world without giving the slightest attention to the ever-varying exhibitions which are presented to them. The sun rises and sets, the seasons change, the heavens every night present new aspects, but these to them are matters of course; they excite no interest, and it is only when some extraordinary phenomenon occurs, such as the blazing comet or the startling earthquake, that their attention is arrested. Another requisite is the power of the perception of truth, which enables the observer to recognize and define with unerring accuracy what he has seen without any tinge of color from *à priori* conceptions. Still another is the faculty of eliminating accidental conditions from those which are essential: and further, the characteristic of perseverance is indispensable.

The fields of scientific labor may be divided into two classes, viz.: those which relate to the empirical observation of facts, and those which refer to the systematic series of investigations relative to the law or cause of special phenomena. As illustrations of the first class are the facts of the phenomena of the physics of the globe, those of ordinary meteorology and natural history; while, as examples of the second, we have the phenomena of chemistry, physics, and astronomy.

The remarks which I have previously made relate principally to the former. In order to elucidate the method of investigation in the latter case, I will suppose the existence of a new phenomenon which is unconnected with any of the present generalizations of science, but of which it is desired to discover the law or the facts with which it is associated. Such facts standing alone form no part of science; they are usually discovered in the course of investigations, and are of great importance in pointing out fields of new research which promise an abundant harvest.

The first step in the investigation is to reproduce the phenomenon; the next, is to form in the mind a provisional hypothesis as to its cause, and in the choice of this we are governed by analogy. For example, if it appears to resemble some of the phenomena of electricity, we *assume* that it is produced by electricity; we next endeavor to ascertain by what known action of electricity such an effect could possibly be produced; for this purpose we invent an hypothesis, or imagine some peculiar action of electricity sufficient to produce the effect in question: we then say to ourselves, if this be true, it will logically follow that a specific result will follow if we make a certain experiment: the experiment is devised and tried, but no positive result is obtained. In order to this negative result, the logical deductions must have been in error, or the experiment must have been defective, or the hypothesis itself erroneous.

We examine each of the two former steps, and finding nothing amiss in them, we conclude that the hypothesis was not true: another hypothesis is then invented, another deduction inferred, and another experiment made; still no result is obtained. At this stage of the research the inexperienced investigator is prone to abandon the pursuit; not so he who has successfully attempted to penetrate the secrets of nature. Undeterred by failure, he changes from time to time his hypotheses, makes new guesses, and again repeats the question as to their truth by means of experiment, until at length nature, as if wearied by his solicitations, grants him a new and positive result; he has now two facts, and an hypothesis to explain them; from this hypothesis he makes a new deduction, which is also tested by a new experiment; but now, perhaps, he obtains a result, which, although of a positive character, is not what he expected. He has, however, made an advance; he has three facts and an hypothesis to explain two of them. In this case he does not usually abandon his preconceived idea, but modifies it until it includes the new fact. With the hypothesis thus improved, he deduces, it may be in rapid succession, a number of new conclusions, the truth of all of which is borne out by the results of the experiments. The investigator now feels that he is on the right track; that the thread of Dedalus is in his hand, and that he will soon be in the full light of day, but usually the escape from the labyrinth is not so easy. In the height of his successful career, it not unfrequently happens that a result is obtained diametrically opposed to his previous generalization, which conclusively forces upon his mind the conviction that he is still far from attaining his end; that he has not yet seized upon the fundamental principle of the phenomena, which have grown into a class under his hands.

At this stage of the inquiry, his self-esteem is much depressed, he throws aside for a while his apparatus, refers to his library for new suggestions; the subject, however, is not discharged from his mind; it still goes with him, and is perpetually recurring; it is mingled with his dreams, and is seen associated with the every-day occurrences of life, until at length, in some happy moment of inspiration, it may be after refreshing sleep, the truth flashes upon him: he catches a more extended conception of the relations of the phenomena; a more comprehensive hypothesis is suggested, from which he is enabled to deduce in succession a large number of new conclusions to be submitted to the test of experiments. These are all found to yield the expected results, and the generalization which has thus been obtained is more than an hypothesis. It is entitled to the name of a verified theory. The investigator now feels amply rewarded for all his toil, and is conscious of the pleasure of the self-appreciation which flows from having been initiated into the secrets of nature, and allowed the place not merely of an humble worshipper in the vestibule of the temple of science, but an officiating priest at the altar.

In this sketch which I have given of a successful investigation, it will be observed that several faculties of the mind are called into operation. First, the imagination, which calls forth the forms of things unseen and gives them a local habitation, must be active in presenting to the mind's eye a definite conception of the modes of operation of the forces in nature sufficient to produce the phenomena in question. Second, the logical power must be trained in order to deduce from the assumed premises the conclusions necessary to test the truth of the assumption in the form of an experiment, and again the ingenuity must be taxed to invent the experiment or to bring about the arrangement of apparatus adapted to test the conclusions.

These faculties of the mind may all be much improved and strengthened by practice. The most important requisite, however, to scientific investigations of this character, is a mind well stored with clear conceptions of scientific generalizations and possessed of sagacity in tracing analogies and devising hypotheses.

Without the use of hypotheses or antecedent probabilities, as a general rule no extended series of investigations can be made as to the approximate cause of casual phenomena. They require to be used, however, with great care, lest they become false guides which lead to error rather than to truth.

It is not enough for a physical investigation that we have the simple idea, which may be embodied in a mathematical equation, —we must see clearly, with the mind's eye, the operations in nature, and how the phenomena are produced in accordance with the well-known laws of force and motion.

As an illustration of what I have said, as well as an original scientific communication, I may be allowed to present in this connection an account of some observations in which I have been engaged during the past summer; and which are an extension of the investigation of the phenomena of sound in its application to fog-signals, of the progress of which I have given an account at different times to the Society.

This year, my attention was called to a special phenomenon

observed for several years past on the coast of Maine, which has been classed among those to which the term "abnormal phenomena of sound" is applied. In August, 1873, this was partially examined, and the result published in the Light-House Report for 1874. In order to investigate it further, I associated myself with Gen. J. C. Duane, Engineer of the 1st Light-House District; Commander H. F. Picking, Inspector of the same District; Mr. Edward L. Woodruff, Assist. Engineer of the 3d District; and Mr. Chas. Edwards, Asst. Engineer of the 1st District.

The phenomenon to be investigated was exhibited in connection with the fog-signal at a station called *White-Head*, on the coast of Maine, at the entrance of Penobscot Bay; it was reported as having been frequently observed by the captains of the steamers plying between Boston and New Brunswick, and had also been witnessed on two different occasions by officers of the Light-House Establishment.

The phenomenon, as reported by these authorities, consisted in hearing the sound of a ten-inch whistle distinctly as the station is approached, till within the distance of from four to six miles, then losing it through a space of about three miles, and not hearing it again until within about a quarter of a mile of the instrument, when it suddenly becomes audible almost in its full power.

This phenomenon, according to the statement of the keeper of the light-house station, is noticed whenever the vessel is approaching the station from the southwest, and the wind is in the same direction. It is especially observed during a fog, when the warning of the signal is most wanted, and which in this locality is always accompanied by a wind from the south or southwest.

Our first object was to verify the phenomenon, and for this purpose we steamed to the southwest, directly against the wind, which was blowing at the time at about the velocity of ten miles per hour; this fortunately happened to be the direction of the wind during which the phenomenon was most frequently observed. The whistle was sounded every minute by an automatic arrangement, and the time at which the several blasts were given could be noted from the vessel by the puffs of steam emitted by the whistle. As we increased our distance from the signal, the sound very slightly diminished in loudness, until the distance was about a half mile, when it suddenly ceased to be heard, and continued inaudible for about a mile farther, when it was faintly heard and continued to increase in loudness until we reached the distance of four miles; at this point it was heard with such clearness that the position of the station could be readily located in the densest fog, but on proceeding still farther in the same direction it gradually diminished and was finally again lost.

As a second experiment we retraced the same line back to the station, and observed the same phenomena in a reversed order. The sound was heard the loudest at a point about four miles from the station, and after that diminished and became inaudible through a space of about two miles, and then suddenly burst forth nearly in full intensity at a distance of a quarter of a mile, and continued loud until the station was reached.

Now, for the investigation of this phenomenon, we may assume provisionally that it is due to a peculiar condition of the atmosphere, either as to heat, pressure, or moisture, or a combination of all of them, which existed at the time in that part of the track of the steamer which may be denominated "the region of silence." But if this were true, such a condition of the atmosphere ought to be indicated by ordinary meteorological instruments. To test this, the temperature of the air was noted through the whole space by an ordinary thermometer, and also its pressure by means of an aneroid barometer, but no variation was observed in these instruments in passing through the air along the path of the vessel.

To complete this series of observations, however, the indications of a delicate hygroscope should have been noted. Unfortunately we were not provided with an instrument of this kind; the fact however that the phenomenon was frequently observed during a fog, or while the air is uniformly saturated with moisture, indicates that the phenomenon is not due to a difference of moisture in the region of silence. Indeed, it is sufficient to remember that a wind was blowing at the rate of ten miles an hour, to be convinced that an isolated portion of air could not remain in a fixed position, even for an instant.

Another hypothesis might be assumed,—that the apparent silence was caused by the transverse reflection, in some way, of sound from the shore, but there was nothing in the configuration of the land which favored such an hypothesis. The only explanation which presented itself was that of the upward refraction of sound, an hypothesis which has been found fertile in new results in previous investigations of the same subject. To test this, and to ascertain the dependence of the phenomenon on the wind, the position of the focus or the origin of the sound was changed. For this purpose, the whistle of the steamer was sounded, while a portion of the observing force was placed at the station; by this arrangement it was found that while the vessel, in reference to the sound of the signal at the station, passed through a region of silence, the observers at the station, who gave attention to the sound from the steamer, heard no interruption of the signal. This experiment was repeated each way, going to and coming from the station.

From this result it appears that the sound going *with* the wind was heard from every point on its course, while the sound moving against the wind was suddenly lost at a given point and not recovered again until a distance had been traversed by the vessel of more than a mile. This result was in strict conformity with the theory of refraction;—in the case of the sounds travelling against the wind, the upper part of the wave would usually be more retarded than the lower, and consequently the sound wave would be thrown upward, above the head of the observer. At a given altitude this difference of velocity would cease and by the tendency of sound constantly to spread, the sound wave would again reach the earth.

But, to test this still farther and to show that the locality was not an essential condition of the existence of the interval of silence, the experiment was repeated on the opposite side of the station, so that the sound from the fog-signal would move in the direction of the wind. Part of the observers were placed at the station and the other part remained on board the vessel; both instruments were sounded, the one in the intervals of the sounding of the other.

In this case the sound from the fog-signal was continuous to those on board of the vessel through a distance of over four miles, and could probably have been heard many miles farther, but the progress of the steamer in that direction was stopped by the land.

From the report of the observers at the station it appeared that as the vessel passed into the distance but one blast was heard during its whole course. In this case (as in the preceding experiment of sailing to the southwest) the sound moving against the wind was refracted upward, and as the whistle was but six inches in diameter it did not give sufficient volume to again reach the earth by spreading.

In experiments of previous years the fact has been shown that the sound is heard under certain conditions better when moving against the wind than in the opposite direction. This was notably the case in the experiments made at Sandy Hook in September, 1874, during which a sound from the west was heard *with* the wind about three times as far as a sound from a similar source was heard from the east, or *against* the wind; and, again, the same sound was heard from the west three times as far as from the east after the wind had settled to a calm; and in a third observation the same phenomenon was observed after the wind had increased to a velocity of ten miles an hour from the east. These effects were afterwards shown to be connected with the fact that the wind during the whole day was blowing strongly from the west, and that the apparent changes of the wind were due to currents at the surface, and thus a sufficient explanation was given to the phenomena observed.

It would appear, however, from the investigations of last summer, that the wave of sound, which has been refracted upward, may descend at a greater distance from its origin than even that at which sound moving with the wind can be heard; probably involving a peculiar case of undulating or compound refraction; but this requires further investigation.

Each series of observations gives rise to new questions, and indicates that the subject is one which is rich in new results. Unfortunately, however, the observations can only be made by the aid of steamers; and these, in the Lighthouse service, can only occasionally be employed in the rare intervals of more imperative duties.

In order, however, to collect data for further use in the explication of the phenomena, the Light-keepers at Block Island and Montauk Point (the eastern portion of Long Island) have been directed to blow the fog-signals for an hour on every Monday morning, each noting whether he can hear the sound from the other station; observing, at the same time, the direction of the wind and the apparent motion of the clouds.

From the result of these observations during the year it appears that the clouds give frequent indications of adverse wind currents, and that the number of times that the sound has been heard against the wind is greater than the number of times it has been heard with the wind; a result which, though unexpected, is not in discordance with previous assumptions.

It will be recollected by the Society that I have, in previous years, mentioned a remarkable phenomenon, which I have denominated the "*ocean echo*." This has also been observed by the scientific adviser of the Trinity Board, and is considered by him as the key of all the abnormal phenomena of sound observed, and as a special illustration of the truth of his hypothesis that such abnormal phenomena are produced by invisible clouds of flocculent atmosphere. The phenomenon in question consists in a reverberation in the form of an echo from a point in the verge of the horizon to which the axis of a fog-trumpet is directed.

In regard to this I first adopted the provisional hypothesis that this was produced by a reverberation from the crests of the waves of the ocean, but it having been stated that the same phenomenon is exhibited while the sea is smooth, this assumption must be abandoned, or in some way modified to suit the observed facts. To test the hypothesis of the reverberation being due to a reflection from an invisible cloud on the verge of the horizon, the trumpet of the large syren on Block Island was gradually elevated from a horizontal to a vertical position, and while in this position it was sounded at intervals for several days; but in no case was an echo heard from the zenith, but in every instance an echo was returned from the horizon around its whole circumference.

In another experiment with a vertical trumpet at Little Gull Island, a small cloud, from which a few drops of rain fell on the area of the base of the Lighthouse, passed directly across the zenith, and during this passage no echo was observed from the cloud, although the trumpet directed toward it was sounded several times in succession.

Again, in order to obtain additional facts in regard to the nature of this echo, observation was made from a vessel, by steaming out directly as if into the region of the echo; *i. e.*, in the direction of that point in the horizon from which the echo appeared to emanate.

In this case the loudness of the echo appeared, as we advanced, to gradually diminish, and to spread itself through a much longer arc of the horizon, while the duration of the echo increased in time.

It would follow from this experiment that the echo is not a reflection from a definite surface, since it would then increase in loudness as the surface is approached; but a series of rebounds from points at various distances.

Another fact of great importance in determining the nature of the echo is that derived from the observations of the keeper at Block Island; he has recorded during the observations of a year, every Monday, the length of the continuance of the echo, the state of the weather, the direction of the wind, and the other meteorological data. From which it is found that the echo is always heard with the sound of the syren during a wind in any direction, and of all intensities; but with less duration after the original blast, during the occurrence of a very high wind, than in calmer weather; and, above all, that it is heard equally well during a dense fog, when evidently the air must be homogeneous and saturated in every part with vapor.

From these facts it appears to me conclusive that the reverberation, constituting the *ocean echo*, cannot be due to invisible clouds. The only hypothesis which suggests itself to my mind as a basis of further investigating this subject, is that in the spread or divergency of the sound the direction of the impulse turns through an angle of a little more than 90°, so as to meet the surface even of the smooth ocean in a direction by which it would be reflected to the ear of the observer, making the angle of reflection equal to the angle of incidence; although from the gradual dispersion of sound-beams, the precise equality of these angles is obviously not very important to the result.

On returning from this excursion by the N. Y. Western Railway to the Hudson River at Troy, opportunity was taken to make some observations on the action of sound in the Hoosac tunnel, through which I passed, accompanied by Mr. E. L. Woodruff, on the afternoon of September 7th. Resting at East Windsor near the western outlet, I spent a considerable part of the following day in making an examination of the work. Mr. W. P. Granger, the chief engineer, and Mr. A. W. Locke, his principal assistant, very courteously furnished a hand-car, and cordially proffered every facility for making any desired investigations. This tunnel, as is familiar to most of those present, is nearly five miles long, rising by an easy grade of 26.4 feet to the mile from either mouth to about the middle of the tunnel, where it opens into a vertical ventilating shaft through the rock, of upwards of a thousand feet in height. The top of this shaft opens between two ridges of the Hoosac mountain, which rise respectively some 400 and 700 feet higher. From the middle of the tunnel when entirely clear of smoke, the distant opening at either end appears as a faint star. The darkness seems oppressive; and when a train is passing through, the air becomes so thickly clouded that the glare of torches cannot be seen at more than a dozen feet distance.

It had been constantly observed by those employed in the tunnel, that during the approach of a locomotive at no great distance, and a few minutes afterward, the sound of the engine was very much deadened and obstructed; so much so indeed, as to imperil the workmen engaged in lining the top of the tunnel with a brick arch, who frequently failed to hear the locomotive until it was close upon them. This obscuration of sound was not unnaturally attributed to the dense clouds of smoke constantly emitted by the locomotive: but this explanation can hardly be accepted as the true one, nor the condition noted, as constituting even an appreciable cause of such acoustic opacity. When we reflect that a puff of exhaust-steam at high temperature is ejected at about every four feet of rail traversed by the driving wheels, it is not difficult to realize that in an atmosphere so systematically made heterogeneous, there must be a very great amount of dispersion and absorption of sound waves struggling through such a medium. This has been well illustrated by the striking experiments of the distinguished physicist of the Royal Institution. A very simple method of confirming this explanation, and of eliminating entirely the effect of the smoke, would be the employment of locomotive engines driven by the combustion of coke or of charcoal. This experimental determination of the question did not occur to me till after we had left the tunnel; but on suggesting it to Mr. A. W. Locke, the assistant engineer in charge, he very obligingly undertook the conduct of such an experiment at the earliest convenient opportunity. The result has not yet transpired.

When the tunnel was entirely clear, and a gentle current of air flowing down the central ventilating shaft and out at the two ends (as is usual in the summer season when the external temperature is higher than the internal), it was observed that a prolonged but irregular echo followed any loud noise, such as the sudden shutting down of the lid of a tool-chest. The unequal or somewhat intermittent character of the echo appeared to result from the irregular surface of the rock forming the walls of the tube. A somewhat similar echo is sometimes returned from the dense foliage of trees. It is proper to add that a very perceptible echo was heard from the portion of the tunnel lined with brick. The effect could in neither case be ascribed to any invisible "flocculence," as the air must have been in a very homogeneous condition.

Inasmuch as in such observations the waves of sound are reflected back to the ear from points at a considerable distance from their origin, this being especially true of the ocean-echoes, we are liable to be seriously misled if we rely too confidently on the experiments of the laboratory; and form hasty generalizations from apparent analogies, without carefully considering *all* the meteorological conditions by which the rays of sound may be deflected, distorted, and diverged. It is now well established by numerous observations and experiments—made independently on both sides of the Atlantic, that the lines of acoustic propagation (conveniently called sound-beams) which are sensibly very rectilinear for the distance of a hundred or two hundred feet, and which are thus obedient to the katoptric and dioptric laws of precise focal convergence, by means of solid mirrors and of gaseous lenses, are yet at the distance of a few miles so strangely contorted and aberrant, as seemingly to contradict all the analogies suggested by our experience with the rays of light. It is the accumulation of comparatively slight divergencies continued through many thousands of yards, whether under the influence of constant conditions, or of changing and reversed conditions, which produces such marked anomalies at the distance of five or of ten miles, and which makes their investigation as laborious as it is instructive and important. And not until we have mastered all the conditions affecting the transmission of sound throughout its entire sensible range, and have thus become enabled to predict its true course, and to announce its varying limits of audibility at the earth's surface, under given circumstances, can we be said to have perfected the theory of this most interesting and indispensable agent of communication.

Remarks were made by Mr. Antisell.

Life and Character of Joseph Henry
J.C. Welling
Columbian College (George Washington University)
148th Meeting
October 26, 1878

Welling, J.C. 1880. "Life and Character of Joseph Henry." *Bulletin of the Philosophical Society of Washington* (2): 203-229.

148TH MEETING. OCTOBER 26, 1878.

Vice-President TAYLOR in the Chair.

Fifty members and visitors present.

Mr. J. C. WELLING read the following address on the

LIFE AND CHARACTER OF JOSEPH HENRY,

and Mr. W. B. TAYLOR gave an historical account of

HIS SCIENTIFIC LABORS.

LIFE AND CHARACTER OF JOSEPH HENRY.

JOSEPH HENRY was born in Albany, N. Y., on the 17th of December, 1799. His grandparents on both his father's and mother's side emigrated from Scotland, and landed in this country on the 16th of June, 1775, the day before the battle of Bunker's Hill. At the age of seven or earlier, for what reason is unknown, he went to live with his maternal grandmother, who resided at Galway, in the county of Saratoga, N. Y., and his father having died soon afterwards, he continued to dwell for years under her roof. At Galway he attended the district school, of which one Israel Phelps was the master, and having there learned how "to write and cipher, too," he was placed at the early age of ten in a store kept in the village by a Mr. Broderick. Receiving from his employer every token of kindness, and, indeed, of paternal interest in his welfare, the boy-clerk, already remarkable for his handsome visage, his slender figure, his delicate complexion, and his vivacious temper, became a great favorite with his comrades, who, according to the customs of the village store, were wont to saunter about the door in summer, and to gather round the stove in winter for the interchange of such trivial gossip as pertains to village life. Though released at this time for the half of each day from the duty of waiting in the store, that he might attend the sessions of the common school in the afternoon, it does not appear that he had as yet evinced any taste for books, notwithstanding the fact, as he afterwards recalled, that his young brain was even then troubled at times with the "malady of thought," as he lost himself in the mazes of revery or speculation about God and creation—"those obstinate questionings of sense and outward things," which the philosophical poet of England has described as the natural misgivings of a "creature moving about in worlds not realized." "Delight and liberty," as was natural to a bright boy in the full flush of his animal spirits, still remained the simple creed of his childhood, until one day his pet rabbit escaped from its warren, and ran into an opening in the foundation of the village church. Finding the hole sufficiently large to admit of pushing his person through it, he followed on all fours in eager pursuit of the fugitive, when his eyes were attracted in a certain direction by a glimmer of light, and groping his way towards it, beneath the church, he discovered that it proceeded from a crevice which led into the vestibule of the building, and which opened immediately behind a bookcase that had been placed in the vestibule, as the depository of the village library. Working his way to the front of the bookcase, he found himself in the presence of all the literature stored on its shelves, and on his taking down the first book which struck his eye, it proved to be Brooke's Fool of Quality, a work of fiction in which views of practical life and traits of mystical piety are artfully blended, insomuch that even John Wesley was inclined to except it from the *auto-da-fé*, which, after the manner of the curate and barber in the story of Don Quixote, he would have gladly performed upon the less edifying products of the novel-writing imagination. Poring over the pages of this fascinating volume, young Henry forgot the rabbit in quest of which he had crept beneath the church. It was the first book he had ever read with zest, because it was the first book he had ever read at the impulse of his "own sweet will." Mrs. Browning has told us that we get no good from a book by being ungenerous with it, by calculating profits—"so much help by so much reading."

> ——— "It is rather when
> We gloriously forget ourselves, and plunge
> Soul-forward, head-long, into a book's profound,
> Impassioned for its beauty and salt of truth—
> 'Tis then we get the right good from a book."

Such was the "soul-forward, head-long plunge" which the boyish Henry now first took in the waters of romance, rendered only the sweeter to him, it may be, because, without affront to innocence, they took the flavor of "stolen waters" from the stealth with which they were imbibed. From that time forth he made frequent visits to this library, by the same tortuous and underground passage, reading by preference only works of fiction, the contents of which he retailed to listening comrades around the stove by night, until, in the end, his patron, who shared in his taste for such "light reading," procured for him the right of access to the library in the regular way, and no longer by the narrow fissure in the rear of the bookcase.

At the age of fifteen he left the store of Mr. Broderick in Galway, and, returning to the place of his birth, entered a watchmaker's establishment in Albany, but finding nothing congenial to his taste in the new pursuit, he soon abandoned it. At this time he had formed a strong predilection for the stage. Two or three years before, while living at Galway, he had seen a play for the first time, on the occasion of a casual visit to Albany, and the impression it made upon his mind was as vivid as that left by the perusal of his first novel. He described and re-enacted its scenes for the wonderment of the Galway youth, and now that he was living in Albany, he could give full vent to his new inclination. His spare money was all spent in theatrical amusements, until at length he won his way behind the scenes, and procured admission to the green room, where he learned how to put a play on the boards, and how to produce the illusion of stage effects. In the skill with which he learned thus early to handle the apparatus of the stage, we may discern, perhaps, the first faint prelude of the skill to which he subsequently attained in handling the "levers and screws" with which, according to Goethe, the experimental philosopher seeks to extort from nature the revelation of her mysteries.

Invited at this period of his life to join a private theatrical association in Albany, known by the name of "The Rostrum," the young enthusiast soon distinguished himself among his fellow-members of riper years by the ingenuity of his dramatic combinations, and the felicity of his scenic effects, insomuch that he was made President of the Society. Meanwhile, the watchmaker had left Albany, and young Henry, no longer having the fear of the silversmith's file and crucible before his eyes, was left free to follow the lead of his dramatic tastes and aspirations. He dramatized a tale, and prepared a comedy; both of which were acted by the association. Indeed, so much was he absorbed in this new vocation, that our amateur Roscius seemed, according to all outward appearance, in a fair way of making a place for himself among the "periwig-pated fellows who tear a passion to tatters" on the stage; or, at the best, of taking rank with the great dramatic artists who, standing in front of the garish foot-lights, "hold the mirror up to nature," in a sense far different from that of the experimental philosopher, standing in the clear beams of that *lumen siccum* which Bacon has praised as the light that is best of all for the eyes of the mind. But in the midst of these "unintelligible dumb shows," under which the unique and original genius of Henry had thus far seemed to be masquerading, we have now come to the time when his mind underwent a great transfiguration, which revealed its native brightness, and a transfiguration as sudden as it was great.

Minds richly endowed, if started at first on a wrong track, may sometimes have, it would seem, an intellectual conversion as marked as that moral conversion which is often visible in the lives of great saints. It certainly was so in the case of Henry. Overtaken in the sixteenth year of his age by a slight accident, which detained him for a season within doors, he chanced, in search of mental diversion, to cast his eyes upon a book which a Scotch gentleman, boarding with his mother, had left upon the table in his chamber. It was Dr. Gregory's Lectures on Experimental Philosophy, Astronomy, and Chemistry. It commences with an address to the young reader, in which the author stimulates him to deeper inquiry concerning the familiar objects around him. "You throw a stone," he says, "or shoot an arrow upwards into the air; why does it not go forward in the air, and in the direction you give it? What force is it that presses it down to the earth? Why does flame or smoke always mount upward? You look into a clear well of water, and see your own face and figure, as if painted there; why is this? You are told it is done by reflection of light. But what is reflection of light?" etc. etc. These queries certainly are very far from representing the *prudens quæstio* of Bacon in even its most elementary form, but they opened to the mind of young Henry an entirely "new world of thought and enjoyment." His attention was enchained by this book as it had not been enchained by the fiction of Brooke, or by the phantasmagoria of the drama.* The book did for him

* He soon became so much interested in this book that its owner gave it to him, and in token of the epoch it had marked in his life, Prof. Henry ever afterwards preserved it among the choicest memorials of his boy-

what the spirits did for Faust when they opened his eyes to see the sign of the macrocosm, and summoned him "to unveil the powers of nature lying all around him." Not more effectual was the call which came to St. Augustine, when, as he lay beneath the shadow of the fig-tree, weeping in the bitterness of a contrite soul, he seemed to hear a voice that said to him: "*Tolle, lege; tolle, lege,*" and at the sound of which he turned away forever from the Ten Predicaments of Aristotle, and all the books of the rhetoricians, to follow what seemed to him the "lively oracles of God." No sooner had Henry recovered from his sickness, than, obedient to the new vision of life and duty which had dawned upon him, he summoned his comrades of "the Rostrum" to meet him in conference, formally resigned the office of President, and, in a valedictory address, announced to his associates that, subordinating the pleasures of literature to the acquisition of serious knowledge, he had determined henceforth to consecrate his life to arduous and solid studies.

There are doubtless those who, in the retrospect of Prof. Henry's youth, as contrasted with the rich flower and fruitage of his riper years, will please themselves with curious speculations on what "might have been," if his rabbit had never slipped its inclosure, if there had been no crack in the wall behind the book-case, or if Gregory's Lectures had never fallen in his way at the critical juncture of his life, much as the great mind of Pascal pleased itself with musing how the fate of Europe might have been changed if the Providential grain of sand in Cromwell's tissue had not sent him to a premature grave; or how the whole face of the earth would have been changed if the nose of Cleopatra had been a little shorter than it was, and so had marred the beauty of face which made her, like another Helen, the *teterrima causa*

hood. In the fly-leaf of the book the following memorandum is found, written by Prof. H. in the year 1837: This book, although by no means a profound work, has, under Providence, exerted a remarkable influence on my life. It accidentally fell into my hands when I was about sixteen years old, and was the first book, with the exception of works of fiction, that I ever read with attention. It opened to me a new world of thought and enjoyment, invested things before almost unnoticed with the highest interest, fixed my mind on the study of nature, and caused me to resolve at the time of reading it that I would devote my life to the acquisition of knowledge.—JOSEPH HENRY.

belli for a whole generation. Such fanciful speculations, it seems to me, are well calculated to import into the philosophy of human life, and into the philosophy of human history, a theory of causation which is as superficial as it is false. As honest Horatio says to Hamlet in the play, when the latter proposes to trace the noble dust of Alexander the Great, in imagination, until perchance it may be found stopping a bung-hole, I feel like saying in the presence of such fine-spun speculations, "'Twere to consider too curiously to consider so." The strong intellectual forces which are organic in a great mind, as the strong moral and political forces which are organic in society, do not depend for their evolution, or for their grand cyclical movements, on the casual vicissitudes which ripple the surface of human life and affairs. To argue in this wise is to mistake occasion for cause, and by confounding what is transient and incidental with what is permanent and pervasive, is to make the noblest life, with its destined ends and ways, the mere creature of accident, and is to convert human history, with its great secular developments, into the fortuitous rattle and chance combinations of the kaleidoscope. We may be sure that Henry was too great a man to have lived and died without making his mark on the age in which his lot was cast, whatever should have been the time, place, or circumstance which was to disclose the color and complexion of his destiny. The strong, clear mind, like the crystal, takes its shape and pressure from the play of the constituent forces within it, and is not the sport of casual influences that come from without.

Armed, however, with his new enthusiasm, the nascent philosopher hastened to join a night school in Albany, but soon exhausted the lore of its master. Encountering next a peripatetic teacher of English grammar, he became, under the pedagogue's drill, so versed in the arts of orthography, etymology, syntax, and prosody, that he started out himself on a grammatical tour through the provincial districts of New York, and returning from this first field of his triumphs as a teacher, he entered the Albany Academy (then in charge of Dr. T. Romeyn Beck) as a pupil in its more advanced studies. Meanwhile, in order to "pay his way" in the academy, he sought employment as a teacher in a neighboring district school, this being, as he afterwards was wont to say, the only office he had ever sought in his life; and in this office he succeeded so well that his salary was raised from $8 for the first month to the munificent sum of $15 for the second month of his service! From pupil in the academy and teacher of the district school, he was soon promoted to the rank of assistant in the academy, and henceforward had ample means for the further prosecution of his studies. Leaving the academy, he next accepted the post of private tutor in the family of the Patroon in Albany, Mr. S. Van Rensselaer; and, devoting his leisure hours to the study of the higher mathematics, in conjunction with chemistry, physiology, and anatomy, he at this time purposed to enter the medical profession, and had made some advances in this direction, when he was called, in the year 1826, to embark in a surveying expedition, set on foot under the auspices of the State government of New York, for the purpose of laying out a road through the southern tier of counties in that State. Starting with his men at West Point, and going through the woods to Lake Erie, he acquitted himself so well in this expedition that his friends endeavored to procure for him a permanent appointment as captain of an engineering corps, which it was proposed to create for the prosecution of other internal improvement schemes, but the bill projected for this purpose having fallen through, Mr. Henry again accepted, though with some reluctance, a vacant chair which was offered him in the Albany Academy.

In connection with the duties of this chair, he now commenced a series of original experiments in natural philosophy—the first connected series which had been prosecuted in this country. Dr. Hare, indeed, had already invented the compound blowpipe, as Franklin before him, by his brilliant but desultory labors, had given an immense impulse to the science of electricity; yet none the less is it true that regular and systematic investigations, designed to push forward the boundaries of knowledge, abreast with the scientific workers of Europe, had hardly been attempted at that era in the United States.

The achievements of Henry in this direction soon began to win for him an increase of reputation as well as an increase of knowledge; but in the midst of the fervors which had come to quicken his genius, he was visited by the fancy (or was it a fact?) that a few of the friends who had hitherto supported him in his high ambition were now beginning to look a little less warmly on his aspirations. Suffering from this source the mental depression which was natural to a sensitive spirit, no less remarkable for its modesty than for its merit, he found solace in the friendly words of good cheer and hopefulness addressed to him by Mr. William Dunlap.* While one day making, with Mr. Henry, a trip down the Hudson River, on board the same steamboat, Mr. Dunlap observed in the young teacher's face the marks of sadness, and, on learning its cause, he laid his hand affectionately on Henry's shoulder, and closed some reassuring advice with the prophetic words, "Albany will one day be proud of her son." The presage was destined to be abundantly confirmed. Soon afterwards came the call to Princeton College, and, because of the wider career it opened to him, the call was as grateful to Henry as its acceptance was gratifying to the friends of that institution. And shortly before this promotion, a new happiness had come to crown his life in his marriage to the excellent lady who still survives him.

He entered upon the duties of his new post in the month of November, 1832, and bringing with him a budding reputation, which soon blossomed into the highest scientific fame, he became the pride and ornament of the Princeton Faculty. The prestige of his magnets attracted students from all parts of the country; but the magnetism of the man was better far than any work of his cunning hand or fertile brain. It was in Princeton, as he was afterwards wont to say, that he spent the happiest days of his life, and they were also among the most fruitful in scientific discovery. Leaving the record of his particular achievements at this epoch to be told by Mr. Taylor, who is so well qualified to do them justice, I beg leave only to refer to this period in the career of Prof. Henry as that in which it was my good fortune to come, for the first time, under the personal influence of the great philosophical scholar, who, after being my teacher in science during the days of my college novitiate at Princeton, continued during the whole of his subsequent life to honor me with a friendship which was as much my support in every emergency that called for counsel and guidance, as it was at all times my joy and the crown of my rejoicing.

* This Mr. Dunlap had been the manager of the Park Theatre in New York, and combined with his dramatic vocation the pursuits of literature and the painter's art. He wrote the "History of Arts and Designs in the United States," a work which was esteemed a standard one at the date of its first publication in 1834.

In the year 1847, when Prof. Henry was in the forty-eighth year of his age, he was unanimously elected by the Regents of the Smithsonian Institution as its Secretary, or Director. At that time the institution existed only in name, under the organic act passed by Congress for its incorporation, in order to give effect to the bequest of James Smithson, Esquire, of London, who by his last will and testament had given the whole of his property to the United States, to found at Washington, under the name of the "Smithsonian Institution," an establishment for "the increase and diffusion of knowledge among men." It does not need to be said that Prof. Henry did not seek this appointment. It came to him unsolicited, but it came to him from the Board of Regents, not only by the free choice of its members, but also at the suggestion and with the approval of European men of science, like Sir David Brewster, Faraday, and Arago, as also of American scientific men, like Bache and Silliman and Hare. I well remember to have heard the late Geo. M. Dallas (a member of the constituent Board of Regents, by virtue of his office as Vice-President of the United States), make the remark on a public occasion, immediately after the election of Prof. Henry as Director of the Smithsonian Institution, that the Board had not had the slightest hesitation in tendering the appointment to him "as being peerless among the recognized heads of American science."

At the invitation of the Regents he drew up an outline plan of the Institution, and the plan was adopted by them on the 13th of December, 1847. The members of this Society, living, as they do, beneath the shadow of the great institution to which Smithson worthily gave his name and his estate, but of which Henry was at once the organizing brain and the directing hand from the date of its inception down to the day of his death, do not need that I should sketch for them the theory on which it was projected by its first Secretary, or that I should rehearse in detail the long chronicle of the useful and multiform services which in pursuit of that theory it has rendered to the cause of science and of human progress. And, moreover, in doing so I should here again imprudently trench on the province assigned to my learned colleague. But I may be allowed to portray the method and spirit which he brought to the duties of this exacting post, at least so far as to say that he proved himself as great in administration as he was great in original research; as skilful in directing the scientific labors of others as he was skilful in the conduct of his own. Seizing, as with an intuitive eye, the peculiar genius of an institution which was appointed to "*increase* knowledge" and to "*diffuse*" it "among men," he touched the springs of scientific inquiry at a thousand points in the wide domain of modern thought, and made the results of that inquiry accessible to all, with a catholicity as broad as the civilized world. And the publications of the Smithsonian Institution, valuable as they are, and replete as they are with contributions to human knowledge, represent the least part of his manifold labors in connection with the Institution. His correspondence was immense, covering the whole field of existing knowledge, and ranging, in the persons addressed, from the genuine scientific scholar in all parts of the world, to the last putative discoverer of perpetual motion, or the last embryo mathematician who supposed himself to have squared the circle.

In accepting a post where he was called by virtue of his office to fecundate the labors of other men rather than his own, Prof. Henry distinctly saw that he was renouncing for himself the paths of scientific glory on which he had entered so auspiciously at Albany and Princeton. He once said to me, in one of the self-revealing moods in which he sometimes unbosomed himself to his intimate friends, that in accepting the office of Smithsonian Secretary he was conscious that he had "sacrificed future fame to present reputation." He was in the habit of recalling that Newton had made no discoveries after he was appointed Warden of the Mint in 1695,* and I believe that the remark is historically accurate, unless we should incline with Biot, against the better opinion of Sir David Brewster, to place after that date the "discoveries" which Newton supposed himself to have made in the Scriptural chronology and in the interpretation of the Apocalypse—discoveries which, whenever made, provoked the theological scoff, as they perhaps deserved the theological criticism of

* The effect of the Wardenship on Newton's scientific labors may be seen in the warmth with which he rebuked Flamsteed for purposing to publish, in 1698, the fact that Newton was then engaged on a revision of the Horroxian theory of the moon. Newton wrote: "I do not love to be printed on every occasion, much less to be dunned and teased by foreigners about mathematical things, or to be thought by our own people *to be trifling away my time when I should be about the King's business.*"

the polemical Bishop Warburton. Yet, having convinced himself that it was a duty which he owed to the cause of science, to sink his own personality in the impersonal institution he was called to conduct, Henry never paused for an instant to confer with flesh and blood, but moved "right onward" in the path of duty, with only the more of steadfastness, because he felt that it was for him a path of sacrifice.

How sedulously he strove to maintain the Institution in the high vocation to which he believed it was appointed no less by a sacred regard for the will of its founder than by an intelligent zeal for the promotion of human welfare, is known to you all. And the success with which he resisted all schemes for the impoverishment of the exalted function it was fitted to perform in the service of abstract science, is a tribute at once to his rare executive skill, and to the native force of character which made him a tower of strength against the clamors of popular ignorance and the assaults of charlatanism. Whatever might be the consequences to himself personally, he was determined to magnify *its* vocation and make *it* honorable. And hence I do not permit myself to doubt that during the long period of his administration, as Secretary of the Smithsonian Institution, covering a period of thirty years, he has impressed upon its conduct a definite direction which his successors will be proud to maintain, not simply in reverence for the memory of their illustrious predecessor, but also in grateful recognition of the fruitful works which, in the pursuit of his enlightened plans, will continue to follow him now that he has rested from his labors.

The rest into which he has entered came to him in a green old age, after a life as full of years as it was full of honors. He was not only blest with an old age which was

——— serene and bright,
And lovely as a Lapland night,

but he also had that which, according to the great dramatist, should accompany old age—"As honor, love, obedience, troops of friends." And the manner of his death was in perfect keeping with the manner of his life. Assured for months before the inevitable hour came that his days on earth were numbered, he made no change in his daily official employments, no change in his social and literary diversions. None was needed. Surprise, I learn, has been expressed that in the full prospect of death he should have "talked" so little about it. But the surprise is quite unfounded. Prof. Henry was little in the habit of talking about himself at any time. Yet to his intimate friends he spoke freely and calmly about his approaching end. Two weeks before he died he said to one such, a gentleman from New York, to whom he was strongly attached: "I may die at any moment. I would like to live long enough to complete some things I have undertaken, but I am content to go. I have had a happy life, and I hope I have been able to do some good." In an hour's conversation which I had with him, six days before he died, he referred to the imminence of his death with the same philosophic and Christian composure. And perfectly aware as he was, on the day before he died, and on the day of his death, that he had already entered the Dark Valley, he feared no evil as he looked across it, but, poised in a sweet serenity, preserved his soul in patience, at an equal remove from rapture on the one hand, or anything like dismay on the other. For his friends he had even then the same benignant smile, the same warm pressure of the hand, and the same affable converse as of yore. With the astronomer, Newcomb, he pleasantly and intelligently discoursed about the then recent transit of Mercury—not unheedful of the great transit he was making, but giving heed none the less to every opportunity for the inquiry of truth. Towards the attendants watching around his couch he was as observant as ever of all the "small sweet courtesies" which marked consideration for others rather than for himself even in the supreme moment of his dissolution. The disciples of Socrates recalled with a sort of pathetic wonder at the calm and intrepid spirit of their dying master, that as the chill of the fatal hemlock was stealing towards his heart, he uncovered his face to ask that Crito should acquit him of a small debt he owed to Esculapius; and so in like manner I recall that our beloved chief did not forget in the hour of his last agony to make provision for the due despatch of a letter of courtesy, which on the day before he had promised to a British stranger.

And so in the full possession of all his great mental powers—in his waking hours filled with high thoughts and with a peace which passed all understanding; in his sleep stealing away

> "To dreamful wastes where footless fancies dwell,"

and talking even there of experiments in sound on board the

steamer Mistletoe, or haply taking note of electric charges sent through imaginary wires at his bidding*—the soul of Joseph Henry passed away from the earth which he had blessed and brightened by his presence. He died ten minutes after twelve o'clock, on the 13th of May, 1878.

From these imperfect notes on the Life of Prof. Henry, I pass to consider some of his traits and characteristics as a man.

He was endowed with a physical organization in which the elements that composed it were not only fine and finely mixed, but were cast in a mould remarkable for its symmetry and manly beauty. The perfection of his "outward man" was not unworthy of the "inward man" whom it enshrined, and if, as a church father has phrased it, "the human soul is the true Shechinah," it may none the less be said that these "fleshly nooks" of ours never appear to so much advantage as when, transfigured by this Shechinah, they offer to the informing spirit a temple which is as stately as it is pure. When Dr. Bentley was called to write the epitaph of Cotes (that brilliant scholar of whom Newton said that, "if he had lived we might have known something"), the accomplished master of words thought it not unmeet to record that the fallen Professor, who had been snatched away by a premature death, was only "the more attractive and lovely because the virtues and graces which he joined to the highest repute for learning were embellished by a handsome person." The same tribute of admiration might be paid with equal justice to the revered Professor whose "good gray head" has just vanished from our sight.

The fascination of Prof. Henry's manner was felt by all who came within the range of its influence—by men with whom he daily consorted in business, in college halls, and in the scientific academy; by brilliant women of society who, in his gracious presence, owned the spell of a masculine mind which none the less was feminine in the delicacy of its perceptions and the purity of

* Prof. Henry took great delight in the acoustical researches which, during the closing years of his life, he made at sea, on board the steamer Mistletoe, while it was in electricity that he won his first triumphs as a scientific man. That his first love and last passion in science still filled his thoughts in his dying moments was attested by the words which even then fell from his lips, in sleep.

its sensibilities; by children, who saw in the simplicity of his unspoiled nature a geniality and a kindliness which were akin to their own. A French thinker has said that *à mesure qu'on a plus d'esprit on trouve qu'il y a plus d'hommes originaux.* It was the breadth and catholicity of Henry's intelligence which enabled him to find something unique and characteristic in persons who were flat, stale, and unprofitable to the average mind.

Gifted with a mental constitution which was "feelingly alive to each fine impulse," he possessed a high degree of æsthetic sensibility to the beautiful in nature and in art. It cannot be doubted that a too exclusive addiction to the analytic and microscopic study of nature, at the instance of science, has a tendency to blunt in some minds a delicate perception for the "large livingness" of Nature, considered as a source of poetic and moral inspiration, but no such tendency could be discovered in the intellectual habitudes of Prof. Henry. To a mind long nurtured by arts of close and critical inquiry into the logic of natural law he none the less united a heart which was ever ready to leap with joy at "the wonder and bloom of the world." When on the occasion of his first visit to England, in the year 1837, he was travelling by night in a stage-coach through Salisbury Plain, he hired the driver to stop, while all his fellow-passengers were asleep, that he might have the privilege of inspecting the ruins of Stonehenge, as seen by moonlight, and brought away a weird "sense of mystery" which followed him in all his after life. At a later day, in the year 1870, after visiting the Aar Glacier, the scene of Prof. Agassiz's well-known labors, he crossed over the mountain to the Rhone Valley, until, at a sudden turn of the road, he came full in the presence of the majestic Glacier of the Rhone. For minutes he stood silent and motionless; then, turning to the daughter who stood by his side, he exclaimed, with the tears running down his cheeks: "This is a place to die in. We should go no further."

And as he rejoiced in natural scenery so also was he charmed with the beauties of art, whether as seen in "the well-stained canvas or the featured stone," and hence it was that he felt as much at home in the *atelier* of the painter or sculptor as in the laboratory of the chemist or the apparatus room of the natural philosopher, and exulted as sincerely in the Louvre or the Corcoran Gallery of Art as in the cabinet of the mineralogist or the museum of the naturalist.

He was as remarkable for the simplicity of his nature as for the breadth of his mind and the acumen of his intellect. Those who analyze the nature and charm of simplicity in a great mind suppose themselves to find the secret of both in the fact that simplicity, allied with greatness, works its marvels with a sweet unconsciousness of its own superior excellence, and it works them with this unconsciousness because it is greater than it knows. Talent does what it can. Genius does what it must. And in this respect, as an English writer has said, there is a great analogy between the highest goodness and the highest genius; for under the influence of either, the spirit of man, "whenever it lifts up its head and shakes its locks," may scatter light and splendor around it, without admiring itself or seeking the admiration of others. And it was in this sense that the simplicity of Henry's nature expressed itself in acts of goodness and in acts of high intelligence, with a spontaneity which hid from himself the transcendent virtue and dignity of the work he was doing; and hence all his work was done without the slightest taint of vanity or tarnish of self-complacency.

As might be expected, he was a fervent lover of the best literature. His acquaintance with the English poets was not only wide but intimate. His memory was stored with choice passages, didactic, sentimental, witty, and humorous, which he reproduced at will on occasions when they were apt to his purpose. His familiarity with fiction dated, as we have seen, from early boyhood, and in this fountain of the imagination he continued to find refreshment for the "wear and tear" of the hard and continuous thought to which he was addicted in the philosopher's study. His knowledge of history was accurate, and it was not simply a knowledge of facts, but a knowledge of facts as seen in the logical coherence and rational explanation which make them the basis of historic generalization. The genesis of the Greek civilization was a perpetual object of interest to his speculative mind, as called to deal with the phenomena of Grecian literature, art, philosophy, and polity.

He was a terse and forcible writer. If, as some have said, it is the perfection of style to be colorless, the style of Henry might be likened to the purest amber, which, invisible itself, holds in clear relief every object it envelops. Without having that fluent delivery which, according to the well-known comparison of Dean Swift, is rarely characteristic of the fullest minds, he was none the less a pleasing and effective speaker—the more effective because his words never outran his thought. We loved to think and speak of him as "the Nestor of American Science," and if his speech, like Nestor's, "flowed sweeter than honey," it was due to the excellent quality of the matter rather than to any rhetorical facility of manner.

He was blest with a happy temperament. He recorded in his diary, as a matter of thanksgiving, that through the kindness of Providence he was able to forget what had been painful in his past experiences, and to remember only and enjoy that which had been pleasurable. The same sentiment is expressed in one of his letters. Armed with this sunny temper which, like the dial, "marks only the hours that shine," he was in his family circle a perpetual benediction. And, in turn, he was greatly dependent on his family for the sympathy and watch-care due in a thousand small things to one who never "lost the child-like in the larger mind." His domestic affections were not dwarfed by the exacting nature of his official duties, his public cares, or his scientific vigils. He had none of that solitary grandeur affected by isolated spirits who cannot descend to the tears and smiles of this common world. He was never so happy as when in his home he was communing with wife and children around the family altar. He made them the confidants of all his plans. He rehearsed to them his scientific experiments. He reported to them the record of each day's adventures. He read with them his favorite authors.* He entered with a gleeful spirit

* The following extract from a diary, kept by one of his daughters, is descriptive of his habits under this head: "Had father with us all the evening. I modelled his profile in clay while he read Thomson's Seasons to us. In the earlier part of the evening he seemed restless and depressed, but the influence of the poet drove away the cloud, and then an expression of almost childlike sweetness rested on his lips, singularly in contrast yet beautifully in harmony with the intellect of the brow above."

Or take this extract from the same diary: "We were all up until a late hour, reading poetry with father and mother, father being the reader. He attempted 'Cowper's Grave,' by Mrs. Browning, but was too tender-hearted to finish the reading of it. We then laughed over the Address to the Mummy, soared to heaven with Shelley's Skylark, roamed the forest with Bryant, culled flowers from other poetical fields, and ended with Tam O'Shanter. I took for my task to recite a part of the latter from memory, while father corrected, as if he were 'playing schoolmaster.'"

into all their joys; with a sympathetic heart into all their sorrows. And while thus faithful to the charities of home, he was intensely loyal to his friends, and found in their society the very cordial of life. Gracious to all, he grappled some of them to his heart with hooks of steel. The friendship, fed by a kindred love of elegant letters, which still lends its mellow lustre to the names of Cicero and Atticus, was not more beautiful than the friendship, fed by kindred talents, kindred virtues, and kindred pursuits, which so long united the late Dr. Bache and Professor Henry in the bonds of a sacred brotherhood. And this was but one of the many similar intimacies which came to embellish his long and useful career.

His sense of honor was delicate in the extreme. It was not only that "chastity of honor which feels a stain like a wound," but at the very suggestion of a stain it recoiled as instantly as the index finger of Mr. Edison's tasimeter at the "suspicion" of heat. I met him in 1847, when, soon after his election as Secretary of the Smithsonian Institution, he had just been chosen to succeed Dr. Hare as Professor of Chemistry in the Medical Department of the University of Pennsylvania, at a salary double that which he was to receive in Washington, and with half the year open to free scientific investigation, because free from professorial duties. It was, he said, the post which, of all others, he could have desiderated at that epoch in his scientific life, but his honor, he added, forbade him to entertain, for a moment, the proposition of accepting it after the obligations under which he had come to the interests represented by the Smithsonian Institution. At a later day, after he had entered on his duties in Washington, and found the position environed with many difficulties, Mr. Calhoun came to him, and urged his acceptance of a lucrative chair in a Southern College, using as a ground of appeal the infelicities of his present post, and the prospect of failing at last to realize the high designs he had projected for the management of the Smithsonian Institution. Admitting that it might be greatly to his comfort and advantage at that time to give up the Smithsonian, he declined at once to consider the proposal made to him, on the ground that his "honor was committed to the Institution." Whereupon Mr. Calhoun seized his hand, and exclaimed, "Professor Henry, you are a man after my own heart."

When in 1853, and again in 1867, he was entreated to accept the Presidency of Princeton College, the College of his love, and the scene of his "happiest days," he instantly turned away from the lure, as feeling that he could not love the dear old College so much if he loved not more the honor and duty which bound him to the establishment in Washington, with which, for good or for evil, he had wedded his name and fortune. And in all other concerns, from the greatest to the least, he seemed like one

> "Intent each lurking frailty to disclaim,
> And guard the way of life from all offence,
> Suffered or done."

The "Man of Ross," portrayed by the pencil of Pope, was not more benevolent in heart or act than Professor Henry. His bounty was large and free. The full soul mantled in his eyes at every tale of woe, and the generous hand was quick to obey the charitable impulses of his sympathetic nature. This benevolent spirit ran like a silver cord through the tissue of his life, because it was interwoven in the very warp and woof of his being, and because it was kept in constant exercise. It appeared not only in acts of kindness to the poor and afflicted, but interpenetrated his whole demeanor, and informed all his conduct wherever he could be helpful to a fellow-man. He did good to all as he had opportunity, from "the forlorn and shipwrecked brother," who had already failed in the voyage of life, to the adventurous young mariner who sought his counsel and guidance for the successful launching of his ship from its ways. Many are the young men, who, in all parts of the land, could rise up to-day and call him blessed, for the blessing he brought to them by the kind word spoken and the kind deed done, each in its season.

Unselfishness was a fundamental trait in the character of Professor Henry, and he made the same trait a fundamental one in his conception of the philosopher's high calling. The work of scientific inquiry was with him a labor of love, not simply because he loved the labor, but because he hoped by it to advance the cause of truth and promote the welfare of man. He never dreamed of profiting by any discovery he made. He would not even have his salary increased, so tenaciously did he hold to the Christlike privilege of living among men "as one that serveth." This was a crown which he would let no man take from him. To the

Government he freely gave, in many spheres of public usefulness, all the time he could spare from his official duties. And it was in one of these subsidiary public labors, as Chairman of the Light-House Board, that he contracted, as he believed, the disease which carried him to the grave.

A sense of rectitude presided over all his thoughts and acts. He had so trained his mind to right thinking, and his will to right feeling and right doing, that this absolute rectitude became a part of his intellectual as well as moral nature. Hence in his methods of philosophizing he was incapable of sophistical reasoning. He sat at the feet of nature with as much of candor as of humility, never importing into his observations "the human being's pride," or the "mystical predominance" of an overweening fancy. He was sober in his judgments. He made no hasty generalizations. His mind seemed to turn on "the poles of truth."

I could not dwell with enough of emphasis on this crowning grace of our beloved friend if I should seek to do full justice to my conception of the completeness it gave to his beautiful character. But happily for me I need dwell upon it with only the less of emphasis because it was the quality which, to use a French idiom, "leaped into the eyes" of all who marked his walk and conversation. In the crystal depths of a nature like his, transparent in all directions, we discern as well the felicity as the beauty of that habit of mind which is begotten by the supreme love of Truth for her own sake—a habit which is as much the condition of intellectual earnestness, thoroughness, and veracity in penetrating to the reality of things, as of moral honesty, frankness, sincerity, and truthfulness in dealing with our fellow-men. The great expounder of the Nicomachean Ethics has taught us, and one of our own moralists has amplified the golden thesis,* that high moral virtue implies the habit of "just election" between right and wrong, and that to attain this habit we need at once an intelligence which is impassioned and an appetite which is reflective. And so in like manner all high intellectual virtue implies a habit of just election between truth and error—an election which men make, other things being equal, according to the degree in which their minds are enamored with the beauty of truth, as also in proportion to the degree in which their appetencies for knowl-

* Dr. James H. Thornwell: Discourses on Truth.

edge have been trained to be reflective and cautious against the enticements of error. I never knew a man who strove more earnestly than Henry to make this just election between right and wrong, between truth and error, or who was better equipped with a native faculty for making the wise choice between them. He had brought his whole nature under the dominion of truthfulness.

But while thus eager and honest in the pursuit of truth he had nothing controversial in his temper. It was a favorite doctrine of his that error of opinion could be most successfully combated, not by the negative processes of direct attack, rousing the pride and provoking the contumacy of its adherents, but rather by the affirmative process of teaching, in meekness and love, the truth that is naturally antagonistic to it. The King of Sweden and Norway made him a Knight of St. Olaf, but St. Olaf's thunderous way of propagating Christianity—by battering down the idols of Norway with Thor's own hammer—is not the way that his American votary would have selected. There was nothing iconoclastic in Henry's zeal for truth. He believed that there is in all truth a self-evidencing quality, and a redemptive power which makes it at once a potent and a remedial force in the world. Hence he never descended to any of those controversies which, in the annals of science, have sometimes made the *odium scientificum* a species of hatred quite as distinct, and quite as lively, too, as its more ancient congener, the *odium theologicum*. When once it was sought to *force* a controversy of this kind upon him, and when accusations were made which seemed to affect his personal honor, as well as the genuineness of his scientific claims, he referred the matter for adjudication to the Regents of the Smithsonian. Their investigation and their report dispensed him from the necessity of self-defence. The simple truth was his sufficient buckler. And this equanimity was not simply the result of temperament. It sprang from the largeness of his mind, as well as from the serious view he took of life and duty. He was able to moderate his own opinions, because, in the amplitude of his intellectual powers, he was able to be a moderator of opinions in the scientific world. You all know with what felicity and intellectual sympathy he presided over the deliberations of this Society, composed as it is of independent scientific workers in almost every department of modern research. Alike in the judicial temper of his mind and in the wide range of his acquisitions he was fitted to be, as Dante has said of Aristotle, "the master of those who know."

And this power of his mind to assimilate knowledge of various kinds naturally leads me to speak of his skill in imparting it. He was a most successful educator. He had many other titles of honor or office, but the title of *Professor* seemed to rank them all, for everybody felt that he moved among men like one anointed with the spirit and power of a great teacher. And he had philosophical views of education, extending from its primary forms to its highest culminations—from the discipline of the "doing faculties" in childhood, to the discipline of the "thinking faculties" in youth and manhood. No student of his left the Albany Academy, in the earlier period of his connection with that institution, without being thoroughly drilled in the useful art of handling figures, for then and there he taught the rudimental forms of arithmetic, not so much by theory as by practice. No student of his left Princeton College without being thoroughly drilled in the art of thinking, as applied to scientific problems, for then and there he was called to indoctrinate his pupils in the rationale as well as in the results of the inductive method. And I will venture to add that no intelligent student of his at Princeton ever failed, in after life, to recognize the useful place which hypothesis holds in labors directed to the extension of science, or failed to discriminate between a working hypothesis and a perfected theory.

Pausing for a moment at this stage in the analysis of Professor Henry's mental and moral traits, I cannot omit to portray the effect produced on the observer by the happy combination under which these traits were so grouped and confederated in his person as to be mutual complements of each other. Far more significant than any single quality of his mind, remarkable as some of his qualities were, was the admirable equipoise which kept the forces of his nature from all interference with the normal development of an integral manhood. He was courtly in his manners, but it was the courtliness which springs from "high thoughts seated in a heart of courtesy," and had no trace of affectation or artificiality; he was fastidious in his literary and artistic tastes, but he had none of that dilettantism which is "fine by defect and delicately weak;" he was imbued with a simplicity of heart which left him absolutely without guile, yet he was shrewd to protect himself against the arts of the designing; he was severe in his sense of honor without being censorious; he was benevolent yet inflexibly just; quick in perception yet calm in judgment and patient of labor; tenacious of right without being controversial; benignant in his moral opinions yet never selling the truth; endowed with a strong imagination yet evermore making it the handmaid of his reason; a prince among men yet without the slightest alloy of arrogance in the fine gold of his imperial intellect; in a word, good in all his greatness, he was, at the same time, great in all his goodness. Such are the limitations of human excellence in most of its mortal exhibitions that transcendent powers of mind, or magnificent displays of virtue exerted in a single direction, are often found to owe their "splendid enormity" to what Isaac Taylor has called "the spoliation of some spurned and forgotten qualities," which are sacrificed in the pursuit of a predominant taste, or an overmastering ambition.* The "infirmities of genius" often attest in their subjects the presence of a mental or moral atrophy, which has hindered the full-orbed development of one or more among their mental and moral powers. But in Professor Henry no one quality of mind or heart seemed to be in excess or deficiency as compared with the rest. All were fused together into a compactness of structure and homogeneity of parts which gave to each the strength and grace imparted by an organic union. And hence, while he was great as a philosopher he was greater as a man, for, laying as he did all the services of his scientific life on the altar of a pure, complete, and dignified manhood, we must hold that the altar which sanctified his gifts was greater than even the costliest offerings he laid upon it.

It will not be expected that I should close this paper without referring to the religious life and opinions of Prof. Henry. If in moral height and beauty he stood like the palm tree, tall, erect, and symmetrical, it is because a deep religious faith was the tap-root of his character. He was, on what he conceived to be

* The phrase, as originally applied by Taylor, is descriptive of certain incomplete ethical systems, but it is equally applicable to certain typical exemplifications of human character, in which "the strength and the materials of six parts of morality have been brought together wherewith to construct a seventh part."

rational grounds, a thorough believer in theism. I do not think he would have said, with Bacon, that he "had rather believe all the fables in the Legend, the Talmud, and the Alcoran, than that this universal frame is without a mind," for he would have held that in questions of this kind we should ask not what we would "rather believe," but what seems to be true on the best evidence before us. He was in the habit of saying that, next to the belief in his own existence, was his belief in the existence of other minds like his own, and from these fixed, indisputable points, he reasoned, by analogy, to the conclusion that there is an Almighty Mind pervading the universe. But when from the likeness between this Infinite Mind and the finite minds made in His image, it was sought, by *a priori* logic, or by any preconceived notions of man, to *infer* the methods of the Divine working, or the final causes of things, he suspected at once the intrusive presence of a false, as well as presumptuous, philosophism, and declined to yield his mind an easy prey to its blandishments. To his eyes much of the free and easy teleology, with which an under-wise and not over-reverent sciolism is wont to interpret the Divine counsels and judgments, seemed little better than a Brocken phantom—the grotesque and distorted image of its own authors, projected on mist and cloud, and hence very far from being the inscrutable teleology of Him whose glory it is to conceal a thing, and whose ways are often past finding out, because His understanding is infinite.

As Prof. Henry was a believer in theism, so also was he a believer in Revealed Religion—in Christianity. He had not made a study of systematic, or of dogmatic, theology, as they are taught in the schools, and still less was the interest he took in polemical divinity, but he did have a theology which, for practical life, is worth them all—the theology of a profound religious experience. He was a fresh illustration of Neander's favorite saying: *Pectus facit theologum.* The adaptation of the Christian scheme to the moral wants of the human soul was the palmary proof on which he rested his faith in the superhuman origin of that scheme. The plan had to him the force of a theory which is scientific in its exact conformity to the moral facts it explains, when these facts are properly known and fully understood.

Hence he was little troubled with the modern conflict between science and religion. History, as well as reason and faith, was here his teacher. He saw that the Christian church had already passed through many epochs of transition, and that the friction incident to such transition periods had only brushed away the incrustations of theological error, and heightened the brightness of theological truth. In a world where the different branches and departments of human knowledge are not pushed forward *pari passu*—where "knowledge comes but wisdom lingers"—he held it nothing strange that the scientific man should sometimes be unintelligible to the theologian, and the theologian unintelligible to the scientific man. He believed, with the old Puritan, that "the Lord has more truth yet to break out of His holy word" than the systematic theologian is always ready to admit; and as the humble minister and interpreter of nature, he was certain that the scientific man has much truth to learn of which he is not yet aware. There must needs be fermentation in new thought, as in new wine, but the vintage of the brain, like the vintage of the grape, is only the better for a process which brings impurities to the surface, where they may be scummed off, and settles the lees at the bottom, where they ought to be. It is under the figure of a vintage that Bacon describes the crowning result of a successful inductive process. When this process has been completed in any direction, it remains for a wider critical and reconciling philosophy to bring the other departments of knowledge into logical relation and correspondence with the new outlook that has been gained on nature and its phenomena.

Erasmus tells us in his Praise of Folly, mingling satire with the truth of his criticism, that in order to understand the scholastic theology of his day, it was necessary to spend six-and-thirty years in the study of Aristotle's physics, and of the doctrines of the Scotists. What a purification of method has been wrought in theology since the times of Erasmus! And for that purification the Church is largely indebted to the methodology of modern science, in clearing up the thoughts, and rationalizing the intellectual processes of men. The gain for sound theology is here unspeakable, and amply repays her for the heavy baggage she has dropped by the way at the challenge of science—baggage which only impeded her march, without reinforcing her artillery.

Hence, as a Christian philosopher, Prof. Henry never found it necessary to lower the scientific flag in order to conciliate an obscurantist theology, and he never lowered the Christian flag in order to conciliate those who would erect the scientific standard over more territory than they have conquered. He had none of that spirit which would rather be wrong with Plato than right with anybody else. He wanted to follow wherever truth was in the van. But better than most men, I think, he knew how to discriminate between what a British scholar calls the duty of "following truth wherever it leads us, and the duty of yielding to the immediate pressure of an argument." He saw, as the same writer adds, that for whole generations "the victory of argument may sway backwards and forwards, like the fortune of single battles," but the victory of truth brings in peace, and a peace which comes to stay. He swept the scene of conflict with the field-glass of a commander-in-chief, and did not set up his trophies because of a brilliant skirmish on the picket lines of science. But he believed in the picket line, and rejoiced in every sharpshooter who fought with loyalty to truth in the forefront of the scientific army.

A man of faith, Prof. Henry was a man of prayer. But his views of prayer were perhaps peculiar in their spirituality. There was nothing mechanical or formal in his theory of this religious exercise. He held that it was the duty and privilege of enlightened Christians to live in perpetual communion with the Almighty Spirit, and in this sense to pray without ceasing. Work was worship, if conducted in this temper. He accepted all the appointments of nature and Providence as the expressions of Infinite Wisdom, and so in everything gave thanks.* He believed that

* The "sweet reasonableness" into which he had schooled his temper was manifested by the great trial which befell him in the year 1865, when the Smithsonian building suffered from the ravages of a fire, which destroyed all the letters written down to that date by Prof. Henry, as Smithsonian Secretary, in reply to innumerable questions relating to almost every department of knowledge. Besides, the Annual Report of the Institution in manuscript, nearly ready for the press, a valuable collection of papers on meteorology, with written memoranda of his own to aid in their digest, and countless minutes of scientific researches which he purposed to make, all perished in the flames. Yet he was more concerned about the loss of Bishop Johns's library, which had been entrusted to his care, than about the loss of his own papers and records. Referring to the latter in a letter written to his friend, Dr. Torrey, a few days after the fire, he held the following language: "A few years ago such a calamity would have paralyzed me for future efforts, but in my present view of life I take it as the dispensation of a kind and wise Providence, and trust that it will work to my spiritual advantage."

familiarity with the order of nature and scientific assurance of its uniformity need not and should not tend to extinguish the instinct, or abolish the motives of prayer by seeming to imply its futility, but should rather tend to purify and exalt the objects of prayer. The savage prays to his idol, that he may have success in killing his enemies. The Hottentot whips and worships his fetich in blind but eager quest of some sensual boon, that he may consume it upon his lusts. The prayers of the Vedic Books are the childish prayers of an unspiritual and childish people. "They pray," says Max Müller, "for the playthings of life, for houses and homes, for cows and horses, and they plainly tell the gods that if they will only be kind and gracious they will receive rich offerings in return." And do *we*, asks the critic of comparative religions, we Christians of this nineteenth century, "do we do much otherwise," if regard be had to the quality of *our* petitions? Prof. Henry held that it was both the duty and privilege of enlightened Christians to "do much otherwise," by praying preëminently, if not exclusively, for spiritual blessings. And hence he held that the highest natural philosophy combines with the highest Christian faith to transfer the religious thoughts, feelings, and aspirations of man more and more from things seen to things unseen, and from things temporal to things eternal. This view of his had nothing of quietism or of mysticism in it. Still less was it the expression of an apathetic stoicism. It was only the philosopher's way of praying to the great All-Father, in the spirit of St. Augustine, "*Da quod jubes, et jube quod vis.*"

I have made this reference to the opinions of Prof. Henry on the relations of science to religion, as also on the relations of natural philosophy to prayer, not only for the light they shed on the character of the man, but also for a reason which is peculiar to this Society, and which it may be a matter of interest for you to know. Immediately after his last unanimous election as the president of our Society, he communicated to me his purpose to make the relations of science and religion, as also the true import of prayer, the subject of his annual presidential address. He gave me an outline of the views he intended to submit, and I have here given but a brief *résumé* of them, according to my recollections of the colloquy, which was only one of many similar conferences previously had on the same high themes. He said that it would be, perhaps, the last time he should ever be called to deliver a presidential address before the Society he so much loved, and that he wished to speak as became an humble patron of Science, believing fully in her high mission, and at the same time as an humble Christian, believing fully in the fundamental truths of Revelation. That he was not able to fulfil this purpose will be as much a source of regret to you as it is to me, but when I compare the valediction which it was in his heart to utter, with the peaceful end which came a few months later to crown his days with the halo of a finished life, I console myself with the thought that no last words of his were needed to seal on our hearts the lesson taught by his long and splendid career. Being dead he yet speaketh.

It is, indeed, the shadow of a great affliction which his death has cast upon our Society, but the light of his life pierces through the darkness, and irradiates for us all the paths of duty and labor, of honor and purity, of truth and righteousness, in which he walked with an eye that never blenched and a foot that never faltered. We shall not see his face any more, beaming with gladness and with the mild splendor of chastened intellect, but we shall feel his spiritual presence whenever we meet in this hall. We shall never hear his voice again, but its clear and gentle tones, as from yonder chair he expounded to us the mysteries of nature, will re-echo in the chambers of memory with only a deeper import, now that he has gone to join the "dead but sceptred sovereigns who still rule our spirits from their urns."

"The origin and early days of the Philosophical Society of Washington"
789th Meeting
April 28, 1917

William H. Dall
Naturalist and Malacologist
National Museum
(The last surviving founding member)

Dall, William H. 1918. "The origin and early days of the Philosophical Society of Washington." *Journal of the Washington Academy of Sciences* 8 (2): 29-34.

Distinguished Men. "I suppose no Society of such a small number of persons as this ever had quite so many distinguished men in portion to the whole number. . . . The Society was made up of men who could say something interesting on almost every branch of science." (Dall 1918, 32)

The Walk to the Smithsonian. "Then General Barnes who was Surgeon-General, and was one of the members, was good enough to offer us more commodious quarters in the city. In those days coming over to the Smithsonian building, especially at night, was something of a task. The paths were not paved; if it happened to be rainy it was a very muddy walk indeed. There was a rather rickety bridge at Tenth Street over a very bad smelling canal which we all had to cross in order to get into the Smithsonian grounds." (Dall 1918, 30)

Learning Outside of Your Field. "The Society was a great boon to all of us who desired to know something of what was going on in the departments of science with which we were not personally acquainted." (Dall 1918, 33)

JOURNAL

OF THE

WASHINGTON ACADEMY OF SCIENCES

Vol. VIII JANUARY 19, 1918 No. 2

HISTORY.—*The origin and early days of the Philosophical Society of Washington.*[1] William H. Dall, National Museum.

The history of the scientific societies in Washington has been admirably told by Mr. G. Brown Goode in his memoir on the origin of the U. S. National Museum. There were, before the formation of the Philosophical Society, two or three societies, all of which finally died. One that included most of the naturalists was called the "Potomac-Side Naturalist's Club;" and it is a matter of some little interest that I had recently a call from Prof. John Chickering, the son of Professor Chickering, of Gallaudet College, who was one of our former members; and he told me that in going over his father's papers, he found the records of meetings of the "Potomac-Side Naturalist's Club."

Then there was the National Institute, which struggled along for a number of years very bravely against adverse circumstances, and finally was obliged to give up on account of the expense of maintaining a museum and other things of that sort which were beyond the means of the members of such a small society.

When I returned from Alaska in 1868, I found that there existed in Washington a club of, I presume, about 20 members, which was, to the best of my recollection, called the "Physical Club." It may be added that its membership comprised some of the most distinguished men of science in Washington, and

[1]An address delivered at the 789th meeting of the Philosophical Society of Washington, April 28, 1917.

Professor Henry was chairman. General Sherman, Admiral Jenkins, J. E. Hilgard, and a number of other men whose names are national property were members of this club. The general method was to have an address or paper by some member of the club, and afterward a social meeting with refreshments furnished by the host of that evening. They were most enjoyable evenings. However, in the course of time, it began to be felt by some of the members that the tax on the less wealthy members of the club was too great. The meetings were held fortnightly, and in the course of the season they would come around several times to the same member. There were others who wished very much to join, but could hardly be accommodated in the houses of the old members; and after more or less discussion about it Professor Henry suggested to some of the men who brought the matter to his attention that they should appoint a committee to organize a society and to have the whole subject laid before the club, to form an organization that would omit the refreshment part of the entertainment; that would make for scientific purposes; and that would be available for any scientific man, either visitor or resident of Washington, and would be restricted to men of science.

The result of this was that a committee was formed whose report you have heard read by the Secretary. The meeting was held in the Regent's room of the Smithsonian Institution, and Professor Henry, by unanimous vote, was made chairman.

A skeleton of a constitution and by-laws, which had been prepared by the Committee of the club, was presented at the meeting and adopted with some amendments. Then General Barnes, who was Surgeon-General, and was one of the members, was good enough to offer us more commodious quarters in the city. In those days coming over to the Smithsonian building, especially at night, was something of a task. The paths were not paved; if it happened to be rainy it was a very muddy walk indeed. There was a rather rickety bridge at Tenth Street over a very bad smelling canal which we all had to cross in order to get into the Smithsonian grounds. I do not know whether any of the present members know that that part of Washington was formerly known and is still known to old residents as "The Island" because it was separated from the city by the James Creek Canal. It is that broad road just south of the Center Market that was formerly the location of this canal, which formed a sort of semicircle and came around the museum grounds enclosing the more elevated land on which the Smithsonian Institution stands, and then went southward through the wide lowland nearly parallel with New Jersey Avenue, Southwest, to rejoin the Potomac, where a small remnant not yet filled in still exists. At that time nearly all of the members of the Society lived in the city and therefore found it desirable to have the place of meeting where they would not have to go through the Smithsonian grounds, often through a considerable amount of mud.

Probably those names that were read by the Secretary mean somewhat less to the members of the Society at present than they meant to us in those days, and I have made an analysis of the committee of the founders, which will, perhaps, throw a little light on the subject.

From the Smithsonian Institution there were of course, Prof. Joseph Henry and William B. Taylor, who was a very erudite man and had a considerable part in the activities of the Institution, T. R. Peale, S. F. Baird, Theodore Gill, and myself. From the Geological Survey (there was at that time no National Survey) came Dr. F. V. Hayden; the Signal Service was represented by General A. J. Myer; from the office of the Nautical Almanac came J. H. C. Coffin, whose great work on the *Winds of the Globe* is well known to all meteorologists.

From the Army there were General Sherman, General Benêt, General Humphreys, General George H. Elliott, General Casey, General Parke, and General Meigs, who built the Cabin John Bridge and had a good deal to do with many of our other principal buildings here in the District; from the Army Medical Museum staff and the Medical Department of the Army there were Dr. Woodward, a microscopist of high reputation; Dr. Otis, who was a distinguished anthropologist; Dr. J. S. Billings, to whom we owe the *Index Medicus;* and Dr. J. K. Barnes, who was Surgeon-General of the Army. Then there were Admiral Foote, Admiral

Sands, and Admiral Jenkins, and from the Naval Observatory Asaph Hall, Simon Newcomb, and William Harkness. These men were all distinguished. I suppose no Society of such a small number of persons as this ever had quite so many distinguished men in proportion to the whole number. I have always felt it a very great honor to have been permitted to join with them in calling myself a founder of the Society.

The original number o founders, that is, those whose signatures were on the list for the formation of the Society, was 43, including Professor Henry who proposed the name, Philosophical Society of Washington, giving to the adjective its original meaning implying the inclusion of all branches of science. No list of members was published in the *Bulletin* until 1874 when the number was 128, there having been four deaths of members, but during that time 85 additional members became connected with the Society. Professor Henry presided over the Society until his death in 1878.

The *Bulletin*, which was issued shortly after the formation of the Society, when enough material had accumulated to form a volume, was reprinted at Professor Henry's suggestion, as a volume of the *Miscellaneous Collections of the Smithsonian Institution.* That procedure was continued during Professor Baird's lifetime. The publications were made up, edited, and printed by the Society, the Smithsonian Institution publication being made by the use of stereotype plates. The reprint was not issued until 1888.

Meetings were held at the offices of the Surgeon-General in the old Ford Theater Building, and were extremely interesting. The Society was made up of men who could say something interesting on almost every branch of science.

We had some very remarkable work presented to the Society. We were privileged, I think, to have the first testing of the telephone. Mr. Bell was introduced by one of the members of the Society. A telephone wire and receiver were strung up in the room where we had our meeting and the transmitter was taken off into another room at some distance and each member of the Society was enabled to hear communications that came from the other room. That was before 1876, when the first public exhibition of the telephone took place at the Centennial Exposition at Philadelphia.

Among other things, I remember a paper by Dr. A. F. A. King in which the mosquito theory of the transmission of malaria was fully set forth. Dr. King was one of those who originated the theory that the disease is transmitted by these insects. Of course, the theory required proof, and it was not till a good many years afterward that proof was furnished by contributors from the Medical Corps of the Army and others.

Most of the papers in the early days were intended to be published elsewhere than in the *Bulletin* of the Society. They were read there for the information of the members, and when the *Bulletin* was printed, it would give the title of the paper and state the place in which it was published, and in that way reference could be had from the *Bulletin* to the place of publication of anything that was read before the Society. At first, of course, the pecuniary resources of the Society were not great and it could not afford to publish many papers, but owing to the fact that most of the members were members of the Government staff under one Bureau or another, and that the publication of their results would naturally have to be through Government agencies, the system adopted was fairly satisfactory. The Society was a great boon to all of us who desired to know something of what was going on in the departments of science with which we were not personally acquainted.

I think that there were none of us but derived welcome and interesting information, and added to our store of knowledge from the communications that came from other members in quite different fields of work. I ought perhaps to mention one of the remarkable things that were done by members of the Society at that time. This was the work of Dr. Woodward of the Army Medical Museum in microscopy. He was the first, as far as I know—at all events in this country, and I think the first anywhere—to succeed in getting a diatom photograph of a perfection and size that would reveal, for instance, all the almost invisible, complete and beautiful ornamentation with which it is provided. Dr. Woodward brought forth his new slides and explained the working of the modifications that he made to his microscope in order to produce this work, and then threw on the screen the beautiful figures, sometimes of an almost invisible diatom enlarged to six feet high, showing every detail of its beautiful structure. They were very interesting indeed. In fact I might go on for a long time with reminiscences of what was brought before us; but we are to hear from others of what developed in the Society as it grew larger and larger and the number of scientific men increased, and how bodies of our members formed other societies and gave to them independent lives.

"The Philosophical Society of Washington through a thousand meetings"
Address to the Society on January 18, 1930

William J. Humphreys
Physicist and Atmospheric Researcher
U.S. Weather Bureau

Humphreys, W.J. 1930. "The Philosophical Society of Washington through a thousand meetings." *Journal of the Washington Academy of Sciences* 20 (13): 245-316 (reproduced here through page 253).

New Societies and the Founding of the Cosmos Club. "Our tree, the Philosophical Society, still bears plenteous fruit of the finest quality and in addition to that, it has furnished many scions that have grown into independent trees, until where once there was but a single tree now there are twenty, or more, each sturdy and an annual bearer of copious crops." . . . Still another scion sprang from this original tree. . . After the formal presentation of scientific papers . . . it often happened that many of those present fell into pleasant and profitable discourse on various topics, a discourse that out of regard to a janitor's endurance, or for some other reason, had to be broken off abruptly, or continued on the street or, and often, in some favorite rathskeller. This was the origin of that other scion, sprung from the desire for free and friendly discourse between scholars, and known the world over as the Cosmos Club." (Humphreys 1930, 250-251)

Come for New Knowledge. "Well, the proper time to come is every time, and the most urgent time to come is when the subject under discussion is the one you know least about." (Humphreys 1930, 252)

Meeting Formality. "The presiding officer always, or nearly always, honored the occasion by appearing in evening dress, and so too on many, if not most, occasions did the speakers; though of course it always was what a man said, and not what he wore when he said it, that really counted. . . . Who, we wonder, when a Woodward was presiding or a Newcomb speaking, would have had the temerity to burn, even for a moment, incense on the altar of the goddess Nicotine;

or in any other way mar, or ease, again as our mood may suggest, the formal dignity of the occasion?" (Humphreys 1930, 252)

The Science. "Here, A. B. Johnson, J. C. Welling, and others, told of the anomalies of sound signals, in which Joseph Henry, John Tyndall, and Lord Rayleigh were so greatly interested. Here, Alexander Graham Bell told of his interesting experiments on the photophone. Here, that fascinating shibboleth, isostasy, was first given to the world, April 27, 1889, in an interesting paper on the problems of Physical Geography by C. E. Dutton. Here, isostasy had much of its development under John F. Hayford, and still flourishes under the constant care of William Bowie. Here, in 1896, a decade before there ever was an aeroplane in the sky, Albert F. Zahm presented a paper on skin friction that is a classic in the science of aviation. Here, C. G. Abbott has kept us informed of all the work of himself and others on solar radiation. Here, E. B. Rosa brought to our knowledge the measurements of extreme accuracy which he, N. E. Dorsey, and their colleagues, were making of fundamental electrical units." (Humphreys 1930, 252)

JOURNAL

OF THE

WASHINGTON ACADEMY OF SCIENCES

VOL. 20 JULY 19, 1930 No. 13

GENERAL SCIENCE.—*The Philosophical Society of Washington through a thousand meetings.*[1] W. J. HUMPHREYS, U. S. Weather Bureau.

Tonight, the one thousandth on which we have foregathered to learn of new discoveries in the limitless field of science, we turn aside from discussions of the outer world to a few minutes of self-contemplation—to recall how, as a society, we came to be, and, lest we forget, to relate again a little of our own history since that natal day some 59 years ago.

Before considering ourselves specifically, however, it will be interesting to review briefly our immediate antecedents. From records kindly examined by our fellow member, Mr. F. E. Brasch, Chief of the Smithsonian Division of the Library of Congress, and from other sources, especially Bulletin 101 of the United States National Museum by the late Richard Rathbun, a former president of this Society, it appears that the spirit of philosophical inquiry came to the Nation's Capital almost with its founding. Washington himself had envisaged the establishment here of a great National University, and Joel Barlow writing to Jefferson, then vice president, from Paris on Sept. 15, 1800, urged that Washington's hope be realized in the establishment of an adequately endowed institution for both collecting and disseminating knowledge, and that "the Institution be called the Philosophic Society." Nothing came immediately of this suggestion. In February 1806 Jefferson and Barlow collaborated in drafting a bill for the establishment of a National Academy and University in Washington, but again there were no tangible results. In this connection Jefferson, then President of the American Philosophical Society, said that he wished there might be a Philosophical Society or Academy at the seat of government with affiliated Academies in each state.

[1] Address before the Philosophical Society of Washington, January 18, 1930.

Such keen interest in science and philosophy, shared also by Fulton, Law, Cutbush, Meigs, Adams and others, inevitably led to the forming of societies holding stated meetings and following announced programs. The earliest of these, the Columbian Agricultural Society, was very ephemeral—founded in 1810 and disbanded in 1812—and had many elements of the County Fair, holding exhibits and awarding premiums for things raised on the farm or made in the home. The next in order, and really the first learned society of Washington, the Metropolitan Society, was formed June 15, 1816, but on the adoption of its constitution, Aug. 8, 1816, changed its name to the Columbian Institute for the Promotion of Arts and Sciences. Its activities were largely agricultural and horticultural together with the collection and display of museum specimens. In 1817 the Medical Society of the District of Columbia was formed, as was also (on March 13) the Washington Botanical Society. The latter, devoted mainly to the study of the plants of the District of Columbia, became inactive in three or four years and quietly vanished in 1826. The Columbian Institute, too, soon became moribund, despite the interest of a few faithful spirits, and in 1837 ceased entirely to exist as an active organization. Perhaps its most conspicuous product was the creation and maintenance for nearly 20 years of a botanic garden, at the very place where thirteen years later the present United States Botanic Garden was established.

On May 15, 1840 the National Institution, later changed to National Institute, was organized in Washington in the expectation, it appears, of controlling and using the Smithsonian bequest, not narrowly and selfishly, but in a broad and liberal spirit, the spirit the Smithsonian Institution, though never controlled by this important and influential organization, was thus enabled more easily to adopt. The National Institute held meetings for about 20 years and published its proceedings in the form of Transactions and other papers. It was disbanded near the beginning of the Civil War, and from 1861 to 1871 the only meetings of scientific men in the city were those of the Saturday Club and the Potomac Side Naturalists' Club. Such then, in merest outline, were the careers of our worthy predecessors.

At the end of this time the many learned men then living here, where life again had become normal, were anxious for the benefits to be derived from the regular meetings of a formal organization, and so it came to pass that the Philosophical Society of Washington, embracing all sciences save those, if they be sciences, of speculative thought, had its origin in the following initiatory letter, dated, it appears, March 12, 1871:

Prof. Joseph Henry, LL.D.

The undersigned respectfully request you to preside at a meeting which they propose to hold for the purpose of forming a society, having for its object the free exchange of views on scientific subjects, and the promotion of scientific inquiry among its members.

M. C. Meigs,	F. V. Hayden,
Benjamin Peirce,	J. E. Hilgard,
Theo. Gill,	J. H. Lane,
Peter Parker,	S. F. Baird,
F. B. Meek,	Walter L. Nicholson,
T. R. Peale,	Wm. H. Dall,
Wm. B. Taylor,	B. Franklin Greene,
Chas. A. Schott,	S. V. Benét,
E. B. Elliott,	Horace Capron,
Thomas Antisell,	Thornton A. Jenkins,
J. J. Woodward,	George H. Elliot,
J. S. Billings,	W. T. Sherman,
J. K. Barnes,	George C. Schaeffer,
C. H. Crane,	Thos. Lincoln Casey,
George A. Otis,	Jno. G. Parke,
Albert J. Myer,	B. F. Sands,
A. A. Humphreys,	A. B. Dyer,
Asaph Hall,	J. B. Wheeler,
Simon Newcomb,	A. B. Eaton,
Wm. Harkness,	Elisha Foote,
B. F. Craig,	Salmon P. Chase.
J. H. C. Coffin,	

The Signers of this letter represented, we see, every branch of both the natural and the exact sciences.

The Society was incorporated in the City of Washington, District of Columbia, in 1901. The letter requesting incorporation was dated May 15, 1901, and signed as follows:

Wm. H. Dall, *Founder*, (L.S.)
Theo. Gill, *Founder*, (L.S.)
Simon Newcomb, *Founder*, (L.S.)

Cyrus Adler,	(L.S.)	James H. Gore,	(L.S.)
Marcus Baker,	(L.S.)	John G. Hagen,	(L.S.)
Louis A. Bauer,	(L.S.)	John F. Hayford,	(L.S.)
Frank H. Bigelow,	(L.S.)	Geo. W. Littlehales,	(L.S.)
F. W. Clarke,	(L.S.)	Charles F. Marvin,	(L.S.)
Wm. A. De Caindry,	(L.S.)	H. M. Paul,	(L.S.)
Robert Fletcher,	(L.S.)	J. W. Powell,	(L.S.)
G. K. Gilbert,	(L.S.)	Richard Rathbun,	(L.S.)
Geo. M. Sternberg,	(L.S.)	J. E. Watkins,	(L.S.)
Otto H. Tittmann,	(L.S.)	Charles K. Wead,	(L.S.)
F. W. True,	(L.S.)	Isaac Winston.	(L.S.)
Chas. D. Walcott,	(L.S.)		

This letter was acknowledged before Henry E. Cooper, Notary Public, May 18, 1901, and filed May 20, 1901.

The presidents of the Society, representing various sciences, have been:—

*Joseph Henry	1871–'78
*Simon Newcomb	1878–'80, 1909
*J. J. Woodward	1881
*W. B. Taylor	1882
*J. W. Powell	1883
*J. C. Welling	1884
*Asaph Hall	1885
*J. S. Billings	1886
*Wm. Harkness	1887
*Garrick Mallery	1888
*J. R. Eastman	1889
*C. E. Dutton	1890
*T. C. Mendenhall	1891
*G. K. Gilbert	1892
*G. Brown Goode	1893
*Robert Fletcher	1894
*W. H. Dall	1895
F. W. Clarke	1896
*Marcus Baker	1897
*F. H. Bigelow	1898
O. H. Tittmann	1899
*G. M. Sternberg	1900
*C. D. Walcott	1901
*Richard Rathbun	1902
J. H. Gore	1903
C. F. Marvin	1904
G. W. Littlehales	1905
*Cleveland Abbe	1906
*John F. Hayford	1907
L. A. Bauer	1908
*C. K. Wead	1909
*R. S. Woodward	1910
A. L. Day	1911
*E. B. Rosa	1912
C. G. Abbot	1913
*L. A. Fischer	1914
W. S. Eichelberger	1915
L. J. Briggs	1916
E. Buckingham	1917
G. K. Burgess	1918
W. J. Humphreys	1919
R. B. Sosman	1920
R. L. Faris	1921
E. C. Crittenden	1922
W. P. White	1923
D. L. Hazard	1924
J. A. Fleming	1925
W. Bowie	1926
*J. P. Ault	1927
Paul R. Heyl	1928
L. H. Adams	1929
Walter D. Lambert	1930

* Deceased.

Returning now to the earliest meetings, we have, quoting from Volume I of the Society's Bulletin:

1st Meeting. March 13, 1871.

Prof. Joseph Henry in the Chair.

In response to this call [the letter quoted above of March 12, 1871, to Joseph Henry] a meeting of the subscribers thereto was convened and held at the Smithsonian Institution, in the Regent's room, on Monday, March 13, 1871. The outline of a Constitution was adopted, and under it the following gentlemen, who collectively should constitute a GENERAL COMMITTEE for the transaction of the business of the Society, were elected officers:—

PRESIDENT.

Joseph Henry.

VICE-PRESIDENTS.

M. C. Meigs,
J. E. Hilgard.
Horace Capron,
Wm. B. Taylor.

TREASURER.

Peter Parker.

SECRETARIES.

B. F. Craig, Theodore Gill.

MEMBERS AT LARGE OF THE GENERAL COMMITTEE.

Thomas Antisell, E. B. Elliott,
J. H. C. Coffin, W. T. Sherman,
S. Newcomb, T. L. Casey,
S. F. Baird, T. A. Jenkins,
J. J. Woodward.

The Constitution was then referred to the General Committee for verbal expression; and the Committee was also empowered to propose Rules and By-Laws for the Government of the Society.

2nd Meeting. March 18, 1871.

The President in the Chair.

Professor S. F. Baird communicated to the Society, on behalf of the author, a copy of a memoir entitled—

OFFICIAL REPORT OF THE YELLOWSTONE EXPEDITION OF 1870, BY LIEUT. G. C. DOANE, 2D U. S. CAVALRY

Thus, organized in the city of Washington, District of Columbia, on March 13, 1871, the Philosophical Society promptly began, March 18, 1871, its scientific career, which it has pursued ever since with much of that quiet dignity and singleness of purpose that characterized its founders.

In his anniversary address, November 18, 1871, Professor Henry spoke on the character and object of this society. Several passages are here quoted from this address. They give us first hand information on our subject, and at the same time reveal to us something of the personality of this great character, the most scholarly of gentlemen and the most gentlemanly of scholars. "Man," he says, "is a sympathetic being, and no incentive to mental exertion is more powerful than that which springs from a desire for the approbation of his fellow men; besides this, frequent interchange of ideas and appreciative encouragement are almost essential to the successful prosecution of labors requiring profound thought and continued mental exertion. Hence, it is important that those engaged in similar pursuits should have opportunities for frequent meetings at stated periods."—"Furthermore, a society of this kind becomes a means of instruction to all its members, the knowledge of each becoming, as it were, the knowledge of the whole."

He then discusses the desirability of publishing a Bulletin. This was done with the aid of the Smithsonian Institution, to and including the 378th meeting, Dec. 19, 1891, of the Society as a whole, and the 64th meeting, April 1, 1891, of the mathematical Section. The resulting eleven volumes contain, in addition to lists of members, officers, by-laws, et cetera, the programs of all meetings, abstracts of many papers, and a few articles in full. Four additional volumes of the Bulletin were published by the Society, bringing its records down to and including the year 1910. On April 22, 1911, it was agreed to defer the further publication of the Bulletin and instead thereof to publish to the extent of 70 pages per year, if so much be needed, in the JOURNAL OF THE WASHINGTON ACADEMY OF SCIENCES and to subscribe to this Journal for all members of the Society who may not at the time also be members of the Academy. This arrangement appears to have been reasonably satisfactory and still is in force.

Professor Henry also comments in his first address to the Society on the relatively large number of scientific workers in Washington, and on the extent of instrumental and library facilities then available.

"With so many facilities," he says, "as exist in the city of Washington for the pursuit of science, this Society would be derelict of duty did it fail to materially aid, through communion of thought and concert of action, the advancement of the great cause of human improvement."

In this recognition of the fact that opportunity implies duty Henry was right. Many centuries ago the same idea was most forcefully expressed in the parable of the trees and their fruit, and it is eternally true. Since that time, more than 58 years ago, when Joseph Henry, the first president of our Society, spoke so wisely, the scientific facilities and opportunities in Washington have vastly increased, and with them have equally grown our imperative duty to produce abundant and good results. Our tree, the Philosophical Society, still bears plenteous fruit of the finest quality and in addition to that, it has furnished many scions that have grown into independent trees, until where once there was but a single tree now there are twenty, or more, each sturdy and an annual bearer of copious crops. Indeed the Society early in its career made the way easy for such a spread and independent development by adopting the following as a standing rule—a rule it long retained:

> Sections representing special branches of science may be formed by the General Committee upon the written recommendation of twenty members of the Society.

Under this authority J. E. Hilgard and nineteen other members of the Society requested on Jan. 27, 1883, that a Section in Mathematical

Science be formed, and such a section was organized on March 29, 1883, and for nearly ten years held independent and successful meetings. After a time, however, when the Society had become more and more mathematical through the loss of the anthropologists, 1879; the biologists, 1880; the chemists, 1884; the entomologists, 1884; and the geographers, 1888—when there was left only mathematics and its more or less immediate dependents, physics, astronomy, geodesy and geology—the mathematical Section ceased to exist as such, its last meeting being the 68th, Nov. 30, 1892. In truth the Society and this Section had become so nearly one and the same that independent meetings on their part were no longer desirable. However, that was by no means the end of the establishment of new scientific societies in Washington.

Still another scion sprang from this original tree, a scion that grew mightily and that had something of the properties of all the others as well as characteristics distinctly its own. It early (1878) came into being in this fashion. After the formal presentation of scientific papers had been completed, whether in some office or elsewhere, it often happened that many of those present fell into pleasant and profitable discourse on various topics, a discourse that out of regard to a janitor's endurance, or for some other reason, had to be broken off abruptly, or continued on the street or, and often, in some favorite rathskeller. This was the origin of that other scion, sprung from the desire for free and friendly discourse between scholars, and known the world over as the Cosmos Club.

And the times have changed. In the earlier days of the Society the universe was its province and every science a congenial topic. The knowledge and the interest of the Henrys and the Newcombs was broad and comprehensive; ours is attenuated. The entomologist's father, for instance, knew bugs, he himself knows a bug, while his son may perhaps, know only a particular flea that lives on that bug! Nor is the physicist any exception to this involution or devolution process. Once he knew something of the phenomena about him, and was happy in that knowledge. Today, well, all too frequently he gets into a metaphysical muddle over the eighteen hundredth portion of a hydrogen atom. Of course the electron is a mighty important thing and is fully worthy of all the attention it is receiving, but one does wish that there were enough good physicists to give many other things the attention they also deserve. Often we look at our program, I fear, and say: "Humph, nothing here about the eighteen hundredth portion of the hydrogen atom. One paper is on the proton, in which I am not interested; and the other on the photon, and that bores me. I shall stay at home." Well, the proper time to come is every time, and the most urgent time to come is when the subject under discussion is the one you know least about.

And the times have changed in another way too. Only a quarter of a century ago, a mere yesterday it seems, the meetings of the Society had the air of formal dignity. The presiding officer always, or nearly always, honored the occasion by appearing in evening dress, and so too on many, if not most, occasions did the speakers; though of course it always was what a man said, and not what he wore when he said it, that really counted. Nor was this the only formality we of the older generation recall regretfully, or amusedly, as our mood may determine. Who, we wonder, when a Woodward was presiding or a Newcomb speaking, would have had the temerity to burn, even for a moment, incense on the altar of the goddess Nicotine; or in any other way mar, or ease, again as our mood may suggest, the formal dignity of the occasion? Why, Sirs, we as lief would have gone to a wedding in our shirt sleeves, or puffed a pipe in church!

The activity of the Society is well shown by the fact that to catalogue only the first 310 of its regular meetings, and the first 35 meetings of the Mathematical Section, required 61 double-column octavo pages in fine print. Nor has the Society been any less active latterly than formerly, though more restricted in its scope. Here, A. B. Johnson, J. C. Welling, and others, told of the anomalies of sound signals, in which Joseph Henry, John Tyndall, and Lord Rayleigh were so greatly interested. Here, Alexander Graham Bell told of his interesting experiments on the photophone. Here, that fascinating shibboleth, isostasy, was first given to the world, April 27, 1889, in an interesting paper on the problems of Physical Geography by C. E. Dutton. Here, isostasy had much of its development under John F. Hayford, and still flourishes under the constant care of William Bowie. Here, in 1896, a decade before there ever was an aeroplane in the sky, Albert F. Zahm presented a paper on skin friction that is a classic in the science of aviation. Here, C. G. Abbott has kept us informed of all the work of himself and others on solar radiation. Here, E. B. Rosa brought to our knowledge the measurements of extreme accuracy which he, N. E. Dorsey, and their colleagues, were making of fundamental electrical units.

Here, too, many another outstanding paper was presented, for these are only excellent examples to illustrate the sustained activity of the Society through a thousand meetings, and the earnest of what the next thousand, and the next and next, will be.

IMPORTANT DATES IN THE SOCIETY'S HISTORY

1871, *March 13*, Society founded.

1874, *June 6*, adoption of the rule that in the official records of the Society no title except "Mr." shall be used.

1883, *January 27*, Mathematical Section formed.

1887, *March 26*, 300th meeting of the Society, and its first in the Assembly Hall of the Cosmos Club, H. Street and Madison Place, where it has met ever since, except occasionally, and then usually in conjunction with some other society and for special reasons. For a number of years previous to this date the meetings had been in the library of the Surgeon General's Office, old Ford's Theater, east side of 10th Street, between E and F, N.W.

1893, *February 18*, 400th meeting of the Society, held at the Corcoran, corner Fourteenth and K streets Northwest, and consisting of a banquet, with toasts and reminiscences; reported in Volume 12 of the Bulletin. Eight founders and 81 other members and guests were present.

1899, *April 15*, celebration of the 500th meeting by a dinner at Rauscher's, Connecticut Ave. and L. Street, with 39 persons present.

1901, *May 20*, Society incorporated in the city of Washington, District of Columbia.

1911, *April 22*, decision adopted to defer the further publication of the Bulletin, and to publish in, and subscribe to, the Journal of the Washington Academy of Sciences.

1925, *December 12*, 55th annual meeting, at which the last surviving founder, William H. Dall, spoke on "Some Recollections of the Founding of the Philosophical Society," and James H. Gore and William H. Holmes gave accounts of the early days of the Society. Unfortunately none of these addresses was preserved.

1930, *January 18*, 1000th meeting, as per copy of program below, at which the foregoing historical sketch was read, supplemented by 31 lantern portraits of deceased prominent members of the Society, and followed by interesting reminiscences by Messrs J. H. Gore and C. F. Marvin.

"Origin and Early Days of the Philosophical Society of Washington"
1962

François N. Frenkiel
Physicist
David W. Taylor Naval Ship Research and Development Center

Frenkiel, François N. 1962. "Origin and Early Days of the Philosophical Society of Washington." *Bulletin of the Philosophical Society of Washington* 16: 9-24.

Only Science. "There were no collateral attractions in the shape of refreshments, ladies were not invited. . ." (Frenkiel 1962, 17)

Founding of the Cosmos Club. "The seven members of the first Board of Management of the Cosmos Club which was founded on November 16, 1878, were all members of the Philosophical Society . . . . [T]hey later served as presidents of the Society and presidents of the Club. . . [T]he Cosmos Club . . . voted unanimously to invite all members of the Society to become original members of the Club." (Frenkiel 1962, 17-18)

No Professional Titles. "On June 6, 1874, the Society adopted a rule that a perfect equality of members is recognized through the omission of all official and professional titles during the meetings." (Frenkiel 1962, 18)

Science and Money. "In his presidential address in 1886, J.S. Billings commented that a scientist should not be interested in money and that a salary of five thousand dollars per year is what he believed to be a limit beyond which the scientific spirit risked being demoralized. No doubt Billings would now allow for a higher limit, if only because the portrait of one of the founders of the Society – Salmon P. Chase – appears on the ten thousand dollar bill." (Frenkiel 1962, 23)

Origin and Early Days of The Philosophical Society of Washington

By François N. Frenkiel

Applied Mathematics Laboratory, David Taylor Model Basin

THE Philosophical Society of Washington was founded in 1871, and was incorporated in 1901, in the District of Columbia. The aims for which the Society was incorporated are "the promotion of science, the advancement of learning and the free exchange of views among its members on scientific subjects."

The name of the Society was selected after considerable deliberation. According to the first president of the Society—Joseph Henry—the term "Philosophical" was chosen to indicate "those branches of knowledge that relate to the positive facts and laws of the physical and moral universe," and the term "Washington" denotes that the Society is a local organization and has no special claims because of its location in the national capital. However, this very location has imposed on the Society certain obligations to the scientific community at large of which it has always been conscious. In his annual address read in 1877, Henry states that he has "taken from the first a deep interest in the Society, knowing that it is intimately connected with the intellectual development of the City of Washington, and that it has a reflex influence upon every part of the United States." In the same address Henry refers to the requirements for membership as follows: "While but comparatively few qualifications are necessary for admittance, yet no person is elected who is not supposed to have at least a high appreciation of science; has some familiarity with its principles, and is capable of doing something in the way of promoting the objects of the Association."

The main activity of The Philosophical Society of Washington has been to hold meetings devoted to the discussion of subjects of scientific interest. These meetings were held on alternate Saturday evenings for almost eighty years and since 1950 have been held on alternate Fridays. Several members of the Society recall the memorable 1000th regular meeting, held on January 18, 1930, at which W. J. Humphreys described the origins of the Society and its history. The 1500th meeting, whose main feature was the address "Sixty Years of Quantum Physics" by E. U. Condon, was preceded by an introduction describing the beginnings of the Society. In resuming the publication of the *Bulletin of The Philosophical Society of Washington* it seems fitting to recall again the origin of the Society, its purposes, its traditions, and the scope of its activities. I shall, for that purpose, quote extensively from the addresses and writings of several founders and past-presidents of the Philosophical Society; their words may have been forgotten, but the impact of their ideas on the Society, as it is at present, has remained.

In planning the new seat of government in the District of Columbia, George Washington gave considerable thought to the establishment of a national university and offered, in 1796, fifty shares of Potomac Company (worth more than twenty thousand dollars) for that purpose. Jefferson supported this project and proposed to attract the entire faculty of the University of Geneva to the new university. In vain Madison tried to convince the House of Representatives to accept Washington's donation; nevertheless, a vote to postpone the resolution was carried by thirty-seven to thirty-six votes. This defeat must have been disappointing to George Washington, who bequeathed the fifty shares in his will and stated in his farewell address: "Promote then, as an object of primary importance, institutions for the general diffusion of knowledge."

The poet and diplomat Joel Barlow in writing to Jefferson in 1800 recommended that Washington's ideas be realized by establishing an endowed institution called the "Philosophic Society" to collect and disseminate knowledge. In 1805, Barlow circulated a prospectus proposing the establishment of a national institution, and in the discussion with Jefferson the latter expressed the wish (in 1807) that there might be a "Philosophical Society" or academy in the capital of the

Nation with affiliated academies in each state. Madison again brought up the question of a national university in his presidential messages to Congress in 1810 and 1817. No immediate results came from these efforts; however, the ideas had been planted and did bear some fruit in later years.

The first Washington society with some scientific undertones was the Columbian Agricultural Society. It was founded in 1810, and, after a short history of exhibits of farm products, it disbanded in 1812. The first real learned society of Washington, originally called the Metropolitan Society, then the Columbian Institute for the Promotion of Arts and Science, was founded in 1816 and devoted its activities to agriculture and horticulture; the Columbian Institute ceased to exist in 1837. The Washington Botanical Society also had an ephemeral history, having formed in 1817 and disbanded in 1826. The first professional society in Washington, founded in 1817, was the Medical Society of the District of Columbia, which is the oldest existing professional society in Washington.

While plans for a national university in Washington failed to find support in Congress and the scientific societies were moribund, an Englishman—James Smithson—changed the course of events by a stroke of his pen. In his will, written in 1826, Smithson left half a million dollars "to found at Washington, under the name of the Smithsonian Institution, an Establishment for the increase and diffusion of knowledge among men." It was not surprising that Smithson, a Fellow of the Royal Society, wished to make a bequest to a scientific institution. What is less clear is how he decided to select Washington for its location. It seems quite probable that Smithson and Barlow met either during the latter's stay in London between 1790 and 1792 or in 1811–1812 in Paris when Barlow was sent as minister to France. Barlow and Smithson were on very friendly terms, respectively, with Joseph Priestley and Henry Cavendish, whose scientific interests were the same and who, like Smithson, were members of the Royal Society.

Congressional reaction to Smithson's bequest was similar to its earlier reaction to Washington's offer. However, with the support of John Quincy Adams and against the views of John C. Calhoun, Congress accepted the bequest. The disposal of the 104,960 golden sovereigns and "eight shillings and seven pence wrapped in paper" which were deposited in the Philadelphia mint in 1838 led to a long struggle between those favoring its use for science and those who wished to use it to support literature and to build a great national library. One of the purposes of the National Institution established in 1840 (later called National Institute) was the expectation of taking over the Smithsonian bequest. The Smithsonian Institution was founded in 1846, and its Regents elected Professor Joseph Henry in 1847 as its first Secretary. The National Institute held meetings for about two decades and then disbanded in 1860. During the following ten years the only two groups of scientists who held meetings in Washington were the Potomac-Side Naturalists' Club and the Saturday Club. The latter group was the direct predecessor of The Philosophical Society of Washington.

Sometime before the outbreak of the Civil War several men interested in science formed a small club called the Saturday Club. This club had neither by-laws nor officers except for a secretary and met on Saturday evenings during the winters at the house of one of its members. The meetings were devoted to discussion of scientific questions and were followed by a social meeting with supper and refreshments provided by the host of the evening. In his book, Hugh McCulloch, Secretary of the Treasury under Lincoln, Johnson, and Arthur, informs us that in 1863 the Saturday Club consisted of seventeen members including Joseph Henry; A. D. Bache, superintendent of the Coast Survey; Peter Parker, former medical missionary and minister to China; Simon Newcomb, professor of mathematics of the Naval Observatory; A. A. Humphreys, who commanded a Union division at Gettysburg and later became chief of engineers; and M. C. Meigs, quartermaster general of the Army, who built the Cabin John Bridge and several other government buildings. However, the organization of the club was so loose that it is difficult to determine whether there was a regular membership. McCulloch speaking of the participants in the club meetings says that "all of them were interesting men, all well known to each other, and some of them to the public, by their scientific and literary attainments; there was not one who would not have been distinguished in any literary and scientific club in this country or in any other; there was not a money-worshipper or time server among them all."

Seven years later the club had increased to more than forty members, and some of the members found it difficult to accommodate the participants in their homes and to provide for the refreshments. Joseph Henry recommended that these difficulties be considered by the club and that plans should be made to organize a society which would be available to any scientific man either visitor or resident of Washington. These ideas led the members of the club to address a letter to Joseph Henry inviting him to preside at a meeting of a committee of founders of the new society. In response to the call, a meeting was held in the Regents' Room of the Smithsonian Institution on Monday, March 13, 1871. The outline of a constitution of the Society was adopted and a General Committee was elected for the transaction of the Society business.

An analysis of the committee of founders made by its last surviving member, William H. Dall (who died in 1927), shows that many agencies and organizations were represented on this committee. The Smithsonian Institution was represented by its secretary Joseph Henry, William B. Taylor, the geologist T. R. Peale, and the zoologists S. F. Baird, T. N. Gill, and W. H. Dall. From the Geological Survey there was F. V. Hayden, the Signal Service was represented by General A. J. Myer, and from the Nautical Almanac Office came J. H. C. Coffin. Then there were Admirals E. Foote, B. F. Sands, and T. A. Jenkins, and from the Naval Observatory Asaph Hall, Simon Newcomb, and William Harkness. From the Army there were Generals W. T. Sherman, S. V. Benét, A. A. Humphreys, G. H. Elliott, T. L. Casey, J. G. Parke, and M. C. Meigs; from the Army Medical Museum Staff and the Medical Department of the Army came the microscopist J. J. Woodward, the anthropologist G. A. Otis, J. S. Billings to whom we owe the *Index Medicus*, and the Surgeon General of the Army J. R. Barnes.

General Barnes offered for the next meetings of the new society the use of the annex of the Surgeon General's office which had been located in the old Ford's Theatre since President Lincoln's death. This meeting place was more easily accessible than the Smithsonian Institution, since most members lived downtown. At that time the way to the Smithsonian Building led across a rickety bridge at Tenth Street over a canal whose approach was announced by its emanations, then along unpaved and often muddy paths. Thus, the second meeting, on

Saturday, March 18, 1871, was held in the old Ford's Theatre. During that meeting, Professor S. F. Baird communicated on behalf of the author a paper entitled "Official Report of the Yellowstone Expedition of 1870" by Lieut. G. C. Doane, 2d U. S. Cavalry.

Fig. 1. Old Ford's Theatre, where regular meetings of the Philosophical Society were held from 1871 to 1887. (From a print found in the archives of the Society.)

A fortnight later, on Saturday, April 1, 1871, the Constitution and the Standing Rules of *The Philosophical Society of Washington* were adopted. According to the first Standing Rules the regular meetings of the Society, held on alternate Saturdays at 8 p. m., were to be devoted to the consideration and discussion of scientific subjects, except for the annual meetings, when the conduct of business of the Society was discussed. This rule was applied the same evening, and the meeting proceeded with its scientific program.

In his inaugural address in 1871, Joseph Henry emphasized the importance of limiting the time spent on the necessary business of the Society, saying: "The bane of many societies is the time consumed in details of business and in the discussion of non-essential points relative to their government. Happily, the organization adopted by this Society obviates this evil and secures the devotion of almost every evening exclusively to its legitimate purposes. For the government of men whose object is the advance of *truth*, but few rules are necessary, and these, unlike the laws of the Medes and Persians—expressed in inexorable codes—must consist of simple principles, readily adaptable to all contingencies."

The sixteen years during which the Philosophical Society was meeting in the old Ford's Theatre were to a great extent its formative years. Some selected examples of the programs of those meetings may give a partial idea of the scope of the discussions and their nature during those earlier times. Thus at a meeting held in 1872, Benjamin Peirce talked about the heat of the sun offering some evidence confirming Lane's views on the gaseous constitution of the sun, William Harkness presented a communication on the constitution of the sun's corona, and Joseph Henry reported his observations made on a journey to California. In the eighteen seventies, travel was not as simple as now and reports from trips made by members of the Society were from time to time the subject of short communications. At another meeting a letter was read giving an account of Professor Hayden's progress in the survey of the Western Territories and General W. T. Sherman described his visit to Egypt; at the same meeting, Simon Newcomb reported the progress of the construction of the new telescope for the Naval Observatory.

During a rather active meeting, held in 1874, C. E. Dutton gave his views on the chemistry of the Bessemer process, Elisha Foote commented on some causes that produced rain, then F. M. Endlich read two papers, one referring to the great wall of China and a second on some specimens of meteorites in general. In connection with the latter paper, Joseph Henry made some remarks on the observations of Professor Newton of

Yale College, in regard to meteor trains and the currents in the upper atmosphere. In the course of the discussion, J. E. Hilgard inquired about the observations on the meteor of Christmas Eve, 1873. In answer to this inquiry, Cleveland Abbe described the plans for the work of a committee appointed for that purpose by the Philosophical Society. (An extensive report on the information collected by this Committee was presented at another meeting.) Some remarks having then been made about earthquake phenomena in North Carolina, John Wesley Powell expressed his conviction that Arizona and Nevada had within recent times been the seat of considerable volcanic and seismic activity.

At the 100th meeting of the Society, held on February 12, 1876, Elisha Gray exhibited his "telaphon" by which musical tones could be transmitted over a telegraph wire and responded to at a distance. A year later, Alexander Graham Bell described the construction of his telephone and exhibited its operation. Bell addressed the Society at other occasions, one of his addresses being devoted to the fallacies concerning the deaf and his views regarding the proper instruction of deaf children. This address was the subject of some objections and extensive comments by E. M. Gallaudet, who presented communications on similar subjects at other meetings.

At a meeting on February 2, 1878, Asaph Hall presented the results of his search for satellites of Mars and the eventual discovery of the two now recognized and proposed for the inner satellite the name of Φόβος and for the outer the name Δειμος.

The first clear and complete presentation of the mosquito theory of transmission of malaria was made, in 1883, by A. F. A. King in his address before the Society. Dr. King suggested that the height at which mosquitoes fly or are brought by the wind from the Potomac flats be studied by placing illuminated fly-traps on the Washington monument. (The monument became controversial in 1854, and after a long interruption the work on the monument resumed in 1880.) It should be recalled that Potomac Park was at that time a marsh with assorted mosquitoes, and this part of the city still deserved the description of a "swamp poodle" used by Mrs. Trollope a few years earlier.

In reminiscing about the meetings in the Ford's Theatre, J. S. Billings describes the "entrance up the narrow stairs, often pervaded with a scientific odor from the laboratory on the lower floor—an odor once compared to that of the deluge at low tide—the devious and complicated route from the head of the stairs, past the General Committee room, to the place of meeting; the rather gloomy room, walled in from floor to ceiling with books from whose dingy backs no light was reflected, and yet in its general aspects and surroundings in many respects appropriate to the objects and purposes of the company gathered therein."

"At those meetings," to quote C. E. Dutton, "the attendance was always large for a local scientific society whose routine meetings were held for the sole purpose of reading formal papers. There were no collateral attractions in the shape of refreshments, ladies were not invited, and the members were expected to limit invitations to non-resident visitors who had, if not distinction, at least a recognized standing, as devotees of science or philosophy." Those formal sessions of the Philosophical Society were followed by what was often called the "adjourned meetings" which were held over beer and pretzels at a tavern on D Street near Tenth Street. W. J. Humphreys, a past president of the Society, and of the Cosmos Club, referred to several Washington societies as scions which sprang from the Philosophical Society. In particular he referred to "a scion that grew mightily and that had something of the properties of all others as well as characteristics distinctly its own. It early (1878) came into being in this fashion. After the formal presentation of scientific papers has been completed, whether in some office or elsewhere, it often happened that many of those present fell into pleasant and profitable discourse on various topics, a discourse that out of regard to a janitor's endurance, or for some other reason, had to be broken off abruptly, or continued on the street or, and often, in some favorite rathskeller. This was the origin of that other scion, sprung from the desire for free and friendly discourse between scholars, and known the world over as the Cosmos Club."

The seven members of the first Board of Management of the Cosmos Club, which was founded on November 16, 1878, were all members of the Philosophical Society and, except for one of them who left the Washington area, they later served as presidents of the Society and presidents of the Club.

After some excitement concerning a rumor that the Philosophical Society contemplated its transformation into a club, an emergency meeting of the Cosmos Club held on December 12, 1878, voted unanimously to invite all members of the Society to become original members of the Club. After moving to its own quarters at the southeast corner of H Street and Madison Place (now named Lafayette Square), the Cosmos Club invited the Philosophical Society to hold its meetings in the Assembly Hall of the Club. The 300th meeting of the Philosophical Society on March 26, 1887, the first one held at the Cosmos Club, marked an epoch in the history of the Society. (Except for some special meetings, the meetings of the Society continued to be held at the Club, which moved to new quarters in 1952 and is now located at the corner of Massachusetts and Florida Avenues.)

Fig. 2. The Dolley Madison House (or Wilkes House), home of the Cosmos Club from 1882 to 1952. Regular meetings of The Philosophical Society of Washington have been held in the assembly hall of the Club since 1887. (From a print found in the archives of the Society.)

On June 6, 1874, the Society adopted a rule that a perfect equality of members is recognized through the omission of all official and professional titles during the meetings. Garrick Mallery, referring to this rule, says: "The Society is probably the only non-esoteric body in the United States in which the titles of Judge and General, Professor and Doctor, Governor, Senator, and even Honorable Member are forbidden; the simple and dignified 'Mr.,' the modern form of Magister, being the only address allowed. Perhaps the plethora of titles and the burdensome hierarchical gradation in Washington compelled this measure of relief." The official records of the Society show that this rule has been applied only during the formal meetings and that it did not apply to guests of the Society. Thus, the published articles and the membership directories include titles, while the minutes of the meeting include the titles of the guests of the Society only. Members of the General Committee did question in more recent times whether the rules permit the introduction of Sir George Thomson by his proper name or whether Cecilia Payne-Gaposchkin should be introduced as Professor or as Madame Payne-Gaposchkin. The founders of the Society did not lack respect for professional titles; they simply tried to emphasize the equality of all members in expressing their scientific views without regard to the position they might hold. The rule, by the way, was amended on May 26, 1945, to include "Mrs." and "Miss." It should be noted that even before that date several ladies were members of the Society; it is not clear whether the amendment was required at that time because a lady-member took part in the discussion which resulted in the dilemma leaving the choice between breaking the rules or being impolite to a lady.

The meetings of the Society were conducted with some decorum and still continue to be somewhat formal. In describing the earlier meetings Humphreys says: "the meetings of the Society had the air of formal dignity. The presiding officer always, or nearly always, honored the occasion by appearing in evening dress, and so too on many, if not most, occasions did the speakers; though of course it always was what a man said, and not what he wore when he said it, that really counted." At the occasion of the 400th meeting on February 18, 1893, J. S. Billings referred to the nature of the meetings, saying: "Science is becoming fashionable of late years, and, as a natural consequence, a sort of scientific *demi-monde* has been evolved, but it has not yet been made at home in the hall of the Philosophical Society," and he continues: "We may, in the spirit of scientific investigation, secure good seats to inspect the latest patterns of skirt-dancing and high-kicking in the places where these are something of a specialty, but we don't want them at home." (We may wonder what Billings, who was a past-president of both the Philosophical Society and the Cosmos Club, would think of the more recent Saturday square-dancing group which is using the Cosmos Club Auditorium.)

The regular meetings of the Society are at present followed by a social hour during which beer and cider, cheese and crackers are served. This has now become a tradition which can be related to the "adjourned meetings" of the Saturday Club. Refreshments were not served at the Ford's Theatre nor were they served during the first years of the meetings at the Cosmos Club. Writing about the 400th meeting which was celebrated by a dinner at The Cochran, the *Washington Post* of February 19, 1893, says that the Society held its "399 regular meetings in a staid and philosophical way without descending to the level of gross and material banquets. Her sister societies ate, drank, and were merry, but the grave and reverend philosophers ate not, neither did they drink," and the article further describes the banquet. It may be interesting to note in passing that the *Washington Post* article refers extensively to the contributions of Professors Henry and Newcomb, and to many other scientists, listing their professional titles. These were the times before the *Post*, apparently at the insistence of a local professional group, decided somewhat arbitrarily to omit the use of professional titles for scientists and educators but to retain them for some other professions.

During the first years of the Society the leading themes of the communications were astronomy, geography, physics, and biology. Geology, meteorology, and anthropology gained the floor of the Society during these beginnings owing to the rapid growth of the Geological Survey, the Weather Bureau, and the Bureau of Ethnology. The meetings devoted to the presentation of the address of a retiring president were attended by a larger audience and were often held at the Smithsonian Institution or in the Columbian University (now The George Washington University). The first presidents of the Society often used the opportunity offered to them in presenting the

Annual Presidential Address to express their views about the purposes of the Society and their views on science.

In his presidential address in 1880, Simon Newcomb discussed the relation of scientific methods to social progress, and referring to the very actual question of tariff, Newcomb asks: "Why is it that while in all subjects of physical science we find a general agreement through a wide range of subjects, and doubt commences only where certainty is not attained, yet when we turn to economic subjects we do not find the beginning of an agreement?" and answering his question he says "The Physicist has an exact nomenclature; uses methods of research well adapted to the objects he has in view; pursues his investigations without being attacked by those who wish for different results; and, above all, pursues them only for the purpose of discovering the truth. In economical questions the case is entirely different. Only in rare cases are they studied without at least the suspicion that the student has a preconceived theory to support. If results are attained which oppose any powerful interest, this interest can hire a competing investigator to bring out a different result. So far as the public can see, one man's result is as good as another's and thus the object is as far off as ever. We may be sure that until there is an intelligent and rational public, able to distinguish between the speculations of the charlatan and the researches of the investigator, the present state of things will continue."

In the same address Newcomb refers to what we would call now the "Two Cultures," saying: "It seems to me that the true educational want is ignored both by those who advocate a classical and those who advocate a scientific education. What is really wanted is to train the intellectual powers, and the question ought to be, what is the best method of doing this?" and he concludes: "The one feature of the scientific spirit which outweighs all others in importance is the love of knowledge for its own sake. If by our system of education we can inculcate this sentiment we shall do what is, from a public point of view, worth more than any amount of technical knowledge, because we shall lay the foundation of all knowledge."

The move of the meeting place from the library of the Surgeon General of the Army in the old Ford's Theatre to the Assembly Hall of the Cosmos Club greatly affected the attendance at the meetings. There were, however, other factors which changed the nature of the meetings of the Philosophical Society, the most important of them being the organization of other, more specialized scientific societies which sprang from the Philosophical Society. Thus, the Anthropological Society was organized in 1879, the Biological Society in 1880, and the Chemical Society in 1884. The need for more specialized meetings found its expression in the formation in 1883 of a Mathematical Section within the Philosophical Society which met until 1892 and held sixty-nine meetings. The Philosophical Society, which had started by including all fields of science in its scope, became more closely identified with the physical and mathematical sciences, geophysics, and biophysics, although it has never taken the character of a specialized society.

A joint committee of the Philosophical, Anthropological, and Biological societies in 1882 prepared the first proposal for a federation of the scientific societies in Washington. This proposal was not accepted at that time; however, the cooperation between the Washington societies followed in another form. Following the initiative of the Philosophical Society, in 1883, the local scientific societies adopted the custom of inviting the members of the societies to listen to the annual address of their presidents. In 1882, the Anthropological and Biological societies gave a course of popular lectures in the National Museum, and, in 1885, the Philosophical Society joined in this work. In 1888, a joint commission of the Anthropological, Biological, Chemical, Geographical, and Philosophical Societies started to work on joint directories of these societies. The first such directory (called "red book") appeared in 1889. In 1897, the council of the Geological Society brought up the question of a joint organization of the scientific societies which led to the appointment of a Committee of Conference of the societies and resulted in the foundation of the Washington Academy of Sciences in 1898.

The first volume of the *Bulletin of The Philosophical Society of Washington* was published in 1874. The publication of the *Bulletin* was done with the aid of the Smithsonian Institution, which reprinted the first ten volumes in its *Miscellaneous Collections*. Starting with volume 11 the method of publication changed; quoting again from the *Washington Post* of February 19, 1893: "One of the most important epochs in the history of the society was when in 1888 the method of publishing the papers presented to the society was changed, and instead of collecting the contributions in a bulky volume at the end of the year, each communication was printed in a separate volume. This has had a wonderful effect upon the circulation of the writings of the members, and these small volumes are at a premium in exchange in all parts of the scientific world. The change in method of publication has been followed by some of the other large scientific societies of the country". This method of publication continued until 1913 when volume 15 was completed. The Society articles appeared later in the *Journal of the Washington Academy of Sciences* until that *Journal* modified its policy in 1960 and ceased to be of an archival nature. The *Bulletin* now resumes publication with volume 16 and is printed by Judd and Detweiler who has printed the *Bulletin* since its foundation in 1874.

In 1931, the Society inaugurated the Joseph Henry Lectures to honor its first president, and, in 1952, the Annual Christmas Lectures. The latter are arranged primarily for high school and advanced junior high students. It is interesting to note that one of the reasons for starting the Christmas Lectures was the fact that the General Committee felt that the value of its investments was increasing more than the needs of the Society required. In fact, the Philosophical Society must have been concerned earlier with becoming too rich, since after having annual dues of five dollars for thirty years it reduced them to three dollars in 1904. Only recently the dues were reinstated to the original five dollars. In his presidential address in 1886, J. S. Billings commented that a scientist should not be interested in money and that a salary of five thousand dollars per year is what he believed to be a limit beyond which the scientific spirit risked being demoralized. No doubt Billings would now allow for a higher limit, if only because the portrait of one of the founders of the Society—Salmon P. Chase—appears on the ten thousand dollar bill.

The view that the soundness of a scientific society is not measured by its financial investments but by its contributions to the scientific activities of direct interest to its members and to the scientific community, however obvious, seems to some to be quite unorthodox. It is not uncommon to confuse the purpose of the existence of a scientific society with the means used for achieving this purpose and after spending much energy to produce those means lose sight of the original purpose. The wisdom of the founders of the Philosophical Society and the experience gained by and transmitted to their successors have protected the Society from such a pitfall.

The proper conduct of the business of a society, however small, requires more work than is often realized by its members. In its ninety-one years of existence, the Philosophical Society has not had an office, an executive secretary, or full-time administrative assistance, but it has been fortunate in having the devotion of its members who have contributed the necessary time to the business of the Society. The by-laws of the Society, its organization, and its traditions limit the time spent by the officers and other members of the General Committee on the necessary but less essential details of the Society government and insure the participation of the Committee members in the scientific discussions and their close contact with the membership. Thus, the regular meetings of the General Committee are held immediately preceding the regular meetings of the Society, and the time devoted to the former is considerably smaller than to the latter.

Regular meetings of the Society are now scheduled on Fridays at 8:15 p.m., and members of the General Committee have been watchful to close their deliberations by 8:14 p.m. lest some member, as it often happens, will call them to order if the meeting of the Society does not start on time. And this is as it should be; for however necessary the deliberations of the General Committee may be, the essence of the business of the Society takes place at its scientific meetings. The care with which the Philosophical Society observes its by-laws and its traditions, and the devotion of its members to the purposes for which it was founded explain the uninterrupted history of the successful development of this oldest existing scientific society of Washington.

The Joseph Henry Lectures
Established 1931

Curtis, H.L. and Judson, L.V. "The Joseph Henry Lectureship of the Philosophical Society of Washington." 1931. *Science* 74 (1911): 168.

THE JOSEPH HENRY LECTURESHIP OF THE PHILOSOPHICAL SOCIETY OF WASHINGTON

THE Philosophical Society of Washington, through its General Committee, has decided to establish a lectureship in honor of its first president, Joseph Henry, Such action at this time is particularly appropriate, since 1931 is the centenary of the discovery of electromagnetic induction, a discovery which has brought honor not only to Henry, but also to Faraday. The purpose and scope of the lectureship can best be shown by quoting the report of the special committee which was adopted by the General Committee:

(1) There is at present in the hands of the treasurer a cash balance . . . and the committee is in agreement that . . . a portion of it be spent . . . in some way which will advance the cause of science and reflect credit on the society.

(2) It is proposed that at suitable intervals of time a speaker be selected to address the society on one of the broad aspects of some field of science, the speaker to review the recent developments or present status of a subject included in or related to the physical sciences.

(3) It is further proposed that the complete address be put in form for publication and submitted to the *Journal* of the Washington Academy of Sciences.

(4) It is further proposed that the aforementioned address shall be called the Joseph Henry Lecture in memory of the first president of the Philosophical Society.

(5) It is further proposed that the first Joseph Henry Lecture be delivered before the society in the fall of 1931, and that thereafter the lecture shall be delivered annually before the society in the spring of the year starting in the spring of 1932.

(6) It is further proposed that the expenses of the speaker incidental to a visit to Washington shall be borne by the society and that in addition an honorarium of one hundred dollars ($100) shall be presented to the lecturer at a suitable time during his visit.

(7) The complete arrangements for any one meeting are to be made by a special committee of three who are to be responsible for selecting the speaker and securing the manuscript in form for publication. No member of the committee should have been a member of the immediately preceding similar committee.

The general committee further provided that the special committee for a given year shall be appointed before October 15 of the preceding year. In accordance with this action, the following committees have been appointed:

For the year 1931: C. G. Abbot, L. H. Adams, *chairman*, R. E. Gibson. For the year 1932: L. J. Briggs, *chairman*, J. H. Taylor, F. E. Wright.

H. L. CURTIS,
President
L. V. JUDSON,
Corresponding Secretary

Joseph Henry Speakers
1931 to 2024
Including Publication Information

1931 – Meeting #1,026 – October 24, 1931
"Certain Aspects of Henry's Experiments on Electromagnetic Induction"
Joseph S. Ames
President, Johns Hopkins University

Wait, G. 1932. "1026th Meeting." *Journal of the Washington Academy of Sciences* 22 (10): 284.
Ames, J. 1931. "Certain aspects of Henry's experiments on electromagnetic induction." *Journal of the Washington Academy of Sciences* 21 (21): 493-503.

1932 – Meeting #1,038 – April 23, 1932
"The Work of Joseph Henry in Relation to Applied Science and Engineering"
Arthur E. Kennelly
Professor Emeritus of Electrical Engineering, Harvard University and the Massachusetts Institute of Technology

Wait, G. 1932. "1038th Meeting." *Journal of the Washington Academy of Sciences* 22 (18/19): 526.
Kennelly, A. 1932. "The work of Joseph Henry in relation to applied science and engineering." *Journal of the Washington Academy of Sciences* 22 (11): 293-310.

1933 – Meeting #1,051 – March 11, 1933
"High Voltage"
Karl T. Compton
President, Massachusetts Institute of Technology

Brickwedde F. 1933. "1051st Meeting." *Journal of the Washington Academy of Sciences* 23 (11): 529.
Compton, K. 1933. "High voltage." *Journal of the Washington Academy of Sciences* 23 (6): 177-297.

1934 – Meeting #1,068 – March 31, 1934
"Spinors"
Oswald Veblen
Professor, The Institute for Advanced Study

Brickwedde, F. 1934. "1068th Meeting." *Journal of the Washington Academy of Sciences* 24 (8): 364.
Veblen, O. 1934. "Spinors." *Journal of the Washington Academy of Sciences* 24 (7): 281-290.

1935 – Meeting #1,083 – March 30, 1935
"What is Electricity?"
Paul R. Heyl
National Bureau of Standards

Silsbee, F. 1935. "1083rd Meeting." *Journal of the Washington Academy of Sciences* 25 (10): 462
Heyl, P. 1935. "What is electricity?" *Journal of the Washington Academy of Sciences* 25 (5): 201-220.

1936 – Meeting #1,095 – February 1, 1936
"The Physical Universe"
Herbert Dingle
Imperial College of Science and Technology, London

Hafstad, L. 1936. "1095th Meeting." *Journal of the Washington Academy of Sciences* 26 (11): 479.
Dingle, H. 1936. "The physical universe." *Journal of the Washington Academy of Sciences* 26 (5): 183-195.

1937 – Meeting #1,115 – March 13, 1937
"Fundamentals of Photosynthesis"
James Franck
Johns Hopkins University
1925 Nobel Prize in Physics

McComb, H. 1938. "1115th Meeting." *Journal of the Washington Academy of Sciences* 28 (1): 32.

Frank, James. 1937. "The Fundamentals of Photosynthesis." Journal of the Washington Academy of Sciences 27 (8): 317-329.

1938 – Meeting #1,132 – March 12, 1938
"Symmetry"
Hermann Weyl
Institute for Advanced Study

McComb, H. 1938. "1132nd Meeting." *Journal of the Washington Academy of Sciences* 28 (8): 381.
Weyl, Hermann. 1938. "Symmetry." *Journal of the Washington Academy of Sciences* 28 (6): 253-271.

1939 – Meeting #1,150 – March 25, 1939
"Recent Developments of Cosmic-Ray Investigation"
Thomas H. Johnston
Assistant Director, Bartol Foundation of the Franklin Institute

Seeger, R. 1939. "1150th Meeting." *Journal of the Washington Academy of Sciences* 29 (7): 311.
Johnston, Thomas H. 1939. "Recent Advances in the Investigation of Cosmic-Rays."
Journal of the Washington Academy of Sciences 29 (6): 233-256.

1940 – Meeting #1,165 – March 16, 1940
"Differences in Physical Properties of Isotopes"
Harold C. Urey
Columbia University
1934 Nobel Prize in Chemistry

Seeger, R. 1940. "1165th Meeting." *Journal of the Washington Academy of Sciences* 30 (9): 408.
Urey, Harold. C. 1940. "The Separation of Isotopes by Chemical Means." *Journal of the Washington Academy of Sciences* 30 (7): 277-294.

1941 – March #1,182 – March 29, 1941
"The Constitution of Diffuse Matter in Interstellar Space"
Otto Struve
Professor of Astronomy, University of Chicago

Mohler, F. 1941. "1182nd Meeting." *Journal of the Washington Academy of Sciences* 31 (8): 377.
Struve, Otto. 1941. "The Constitution of Diffuse Matter in Interstellar Space." *Journal of the Washington Academy of Sciences* 31 (6): 217-258.

1942 – Meeting #1,210 – December 19, 1942
"The Scientific Significance of Ferromagnetism"
Francis Bitter
Massachusetts Institute of Technology & Naval Bureau of Ordnance

Blake, A. 1944. "1210th Meeting." *Journal of the Washington Academy of Sciences* 34 (5): 162.
Bitter, Francis. 1943. "The Scientific Significance of Ferromagnetism." *Journal of the Washington Academy of Sciences* 33 (8): 235-238.

1943 – No Henry Lecture

1944 – Meeting #1,234 – April 29, 1944
"Faster than Sound"
Theodore von Karman
California Institute of Technology

Blake, A. 1945. "1234th Meeting." *Journal of the Washington Academy of Sciences* 35 (3): 102.
von Karman, Theodore. 1945. "Faster than Sound." *Journal of the Washington Academy of Sciences* 35 (5): 137-144.

1945 – Meeting #1,249 – April 14, 1945
"Causality, Statistics, and Quantum Mechanics"
John von Neumann
Institute for Advanced Study, Princeton University

Sherman, K. 1945. "1249th Meeting." *Journal of the Washington Academy of Sciences* 35(11): 371.

1946 – Meeting #1,268 –May 25, 1946
"The Pitch, Loudness and Quality of Musical Tones"
Harvey Fletcher
Bell Telephone Laboratories

Cheney, W. 1948. "1268th Meeting." *Journal of the Washington Academy of Sciences* 38 (1): 25.
Fletcher, Harvey. 1946. "The Pitch, Loudness and Quality of Musical Tones: A demonstration lecture introducing the new Tone Synthesizer." *American Journal of Physics* 14 (4): 215-225.

1947 – Meeting #1,277 –February 15, 1947
"High Centrifugal Fields"
Jesse W. Beams
Professor of Physics, University of Virginia

Cheney, W. 1948. "1277th Meeting." *Journal of the Washington Academy of Sciences* 38 (1): 29.
Beams, Jesse W. 1947. "High Centrifugal Fields." *Journal of the Washington Academy of Sciences* 37 (7): 221-241.

1948 – Meeting #1,298 – March 27, 1948
"Cosmic-Ray Mesotrons"
Frits Zernicke
Johns Hopkins University
1953 Nobel Prize in Physics

Cheney, W. 1948. "1298th Meeting." *Journal of the Washington Academy of Sciences* 38 (8): 287.

1949 – Meeting #1,317 – May 14, 1949
"Basement Rocks Beneath the Ocean"
Maurice Ewing
Columbia University

Mahan, A. 1949. "1317th Meeting." *Journal of the Washington Academy of Sciences* 39 (12): 425.

1950 – Meeting #1,332 –May 13, 1950
"New Phenomena of Electronic Conduction in Semiconductors"
William Shockley
Bell Telephone Laboratories
1956 Nobel Prize in Physics

Secretary's Abstract. 1950. "1332nd Meeting." *Journal of the Washington Academy of Sciences* 40 (9): 311-312.

1951 – Meeting #1,344 –March 23, 1951
"The Formation of Stars"
Lyman Spitzer
Princeton University

Wood, L. 1952. "1344th Meeting." *Journal of the Washington Academy of Sciences* 42 (10): 336-337.
Spitzer, Lyman. 1951. "The Birth of Stars from Interstellar Clouds." *Journal of the Washington Academy of Sciences* 41 (10): 309-318.

1952 – Meeting #1,361 –April 18, 1952
"Meteors and Meteorites"
Sydney Chapman
Oxford University

Wood, L. 1952. "1361st Meeting." *Journal of the Washington Academy of Sciences* 42 (10): 343.
Chapman, Sydney. 1952. "Meteors and Meteorites." *Journal of the Washington Academy of Sciences* 42 (9): 273-282.

1953 – Meeting #1,378 –April 24, 1953
"Mesons and Nuclear Forces"
Hans Bethe
Cornell University
1967 Nobel Prize in Physics

1955. "1378th Meeting." *Journal of the Washington Academy of Sciences* 45 (12): 384.
Bethe, Hans. 1954. "Mesons and Nuclear Forces." *Journal of the Washington Academy of Sciences* 44 (4): 97-105.

1954 – Meeting #1,391 –March 26, 1954
"Advantages and Disadvantages of a Causal Interpretation of Quantum Mechanics"
Henry Margenau
Yale University

1956. "1391st Meeting." *Journal of the Washington Academy of Sciences* 46 (1): 28.
Margenau, Henry. 1954. "Advantages and Disadvantages of Various Interpretations of Quantum Theory." *Journal of the Washington Academy of Sciences* 44 (9): 265-276

1955 – Meeting #1,408 – March 25, 1955
"Tritium in Nature"
Willard F. Libby
U.S. Atomic Energy Commission
1960 Nobel Prize in Chemistry

1956. "1408th Meeting." *Journal of the Washington Academy of Sciences* 46 (6): 194.
Libby, Willard F. 1955. "Tritium in Nature." *Journal of the Washington Academy of Sciences* 45 (10): 301-314.

1956
George P. Thomson
1937 Nobel Prize in Physics

1957. "1427th Meeting." *Journal of the Washington Academy of Sciences* 47 (1): 30.
Thomson, G. 1956. "Atomicity and patterns." *Journal of the Washington Academy of Sciences* 46 (7): 201-205.

1957
George E. Kimball

Kimball, G. 1958. "A critique of operations research." *Journal of the Washington Academy of Sciences* 48 (2): 33-37.

1958
Robert H. Dicke

Dicke, R. 1958. "Gravitation an enigma." *Journal of the Washington Academy of Sciences* 48 (7): 213-223.

1959
Cecilia Payne-Gaposchkin

1960
"Elementary Particles"
Tsung-Dao Lee
1957 Nobel Prize in Physics

1961 – Meeting #1,510 –May 12, 1961
"New Developments in High Magnetic Field Research"
Francis Bitter
Massachusetts Institute of Technology

1962. "1510th Meeting." *Bulletin of the Philosophical Society of Washington* 16: 193.
Bitter, F. 1961. "New developments in high-magnetic field research." *Physics Today* 14 (9): 22-28.
Bitter, F. 1961. "New developments in high-magnetic field research." *Bulletin of the Philosophical Society of Washington* 16: 121-136.

1962 – Meeting #1,526 –May 11, 1962
"The Theory of the Electron"
Fritz Rohrlich
State University of Iowa

https://pswscience.org/meeting/the-theory-of-the-electron/

1963
George E. Pake
Washington University in St. Louis

1964
Peter P. Ewald
American Crystallographic Association

1965
Thomas Gold
Cornell University

1966
John Archibald Wheeler
Princeton University

1967
George Porter
Director, Royal Institution of Great Britain
1967 Nobel Prize in Chemistry

1968
Eugene P. Wigner
Princeton University
1963 Nobel Prize in Physics

1969 – Meeting # 1,640 – May 2, 1969
"The Nature of Pulsars"
Frank D. Drake
Cornell University

1970 – Meeting #1,656 – May 1, 1970
"A Review of Evidence in Regard to the Structure of the Moon"
Harold C. Urey
University of California at San Diego
1934 Nobel Prize in Chemistry

1971 – Meeting #1,673 – May 14, 1971
Philip Handler
National Academy of Sciences

1972 – Meeting #1,690 – May 12, 1972
"Neutron Stars and Black Holes in Our Galaxy"
Remo Ruffini
Joseph Henry Physical Laboratories, Princeton University

1973 – Meeting #1,707 – May 4, 1973
"Models of Population Growth, Competition and Diffusion" Elliott W Montroll
University of Rochester

1974 – Meeting #1,723 – May 10, 1974
"Is There Chance?"
Mark Kac
The Rockefeller University

1975
Nicholaas Bloembergen
1981 Nobel Prize in Physics

1976
Nathan Reingold
Smithsonian Institution

1977
Isidor I. Rabi
Columbia University
1944 Nobel Prize in Physics

1978 – April 28, 1978
"Time and Structure"
Ilya Prigogine
1997 Nobel Prize in Chemistry

1979
Banesh Hoffman
Queens College, City University of New York

1980
Philip Leder
Harvard University

1981
Val Fitch
Princeton University
1964 Nobel Prize in Physics

1982 – May 7, 1982
"The Electrochemical Intricacies of Neurobiology" Marshall Nirenberg
National Heart, Lung, and Blood Institute
1968 Nobel Prize in Physiology or Medicine

1983 – Unknown

1984 – Unknown

1985 – Unknown

1986 – Unknown

1987 – Unknown

1988 – Meeting #1,931 – May 20, 1988
Walter Gilbert
Harvard University
1980 Nobel Prize in Chemistry

1989 – Meeting #1,948 – May 26, 1989
"The Public Perception of Genetics"
Maxine Singer
President, Carnegie Institution

1990 – May 18, 1990
"Kuru in Kine and Oravske Kuru"
D. Carleton Gajdusek
National Institutes of Health
1976 Nobel Prize in Physiology or Medicine

1991 – Meeting #1,979 – May 10, 1991
"The Case for Ending Drug Prohibition"
Joe G. Foreman
Scientist and Former President of the Society

1992 – Meeting #1,997 – May 15, 1992
"Is it Possible to Construct a Perception Machine?"
Gerald M. Edelman
Rockefeller University
1972 Nobel Prize in Physiology or Medicine

1993 – Meeting #2,013 – May 14, 1993
"Origin, Evolution, and Distribution of Life in the Universe"
Cyril Ponnamperuma
Director, Laboratory of Chemical Evolution, University of Maryland

1994 – Meeting #2,029 – May 13, 1994
"Environmental Influences on Brain Structure and Function"
Marian Diamond
Professor of Anatomy and Director Lawrence of Science
University of California, Berkeley

1995 – Meeting #2,045 – May 12, 1995
"Designer Brain"
Richard M. Restak
George Washington University Medical School

1996 – Meeting #2,060 – May 3, 1996
"How do Protein Molecules Self-Assemble?"
George D. Rose
Johns Hopkins University School of Medicine

1997 – Meeting #2,075 – May 2, 1997
"Mars in Fact versus Fiction"
Donna Lee Shirley
Jet Propulsion Laboratory

1998 – Meeting #2,091 – May 8, 1998
"What Hubble Didn't Know About the Universe"
Vera C. Rubin
Carnegie Institution of Washington

1999 – Meeting #2,106 – May 7, 1999
"Quantum Crystallography, a Tool for Enhancing Information from Crystallographic Data"
Isabella Karle
Naval Research Laboratory

2000 – Meeting #2,118 – May 12, 2000
"Global Warming: Hot Talk, Cold Science"
S. Fred Singer
President, Science and Environmental Policy Project

2001 – Meeting #2,132 – May 11, 2001
"The Human Computer and the Birth of the Information Age"
David A. Grier
Associate Professor, George Washington University

https://pswscience.org/meeting/the-human-computer-and-the-birth-of-the-information-age/

2002 – Meeting #2,147 – May 17, 2002
"The Cholesterol Myths"
Uffe Ravnskov
Independent Researcher

2003 – Meeting #2,163 – May 16, 2003
"Biology and Humans in Space"
Baruch Blumberg
1976 Nobel Prize in Physiology or Medicine

2004 – Meeting #2,178 – May 28, 2004
Oceans and Human Health
Rita R. Colwell
Retiring Director, National Science Foundation

2005 – Meeting #2,193 – May 13, 2005
"Time, Einstein, and the Coolest Stuff in the Universe"
William Phillips
NIST
1997 Nobel Prize in Physics

2006 – Meeting #2,207 – May 12, 2006
"Latest Scientific Results from the Mars Exploration Rovers"
Jim Garvin
Chief Scientist, NASA Goddard Space Flight Center

2007 – Meeting #2,223 – May 11, 2007
"The Very First Light"
John Mather
Senior Astrophysicist, NASA Goddard Space Flight Center
2006 Nobel Prize in Physics

2008 – Meeting #2,239 – May 16, 2008
"Arsenic in Ground Water: Aquatic Chemistry and Mitigation through a Sustainable Filtration Technology"
Abul Hussam
Associate Professor, Department of Chemistry and Biochemistry, George Mason University

2009 – Meeting #2,255 – May 1, 2009
"Our Microbial Partners and Their Roles in Health and Disease"
Claire M. Fraser-Liggett
Director of the Institute for Genome Sciences, University of Maryland

2010 – Meeting #2,270 – May 14, 2010
"Cosmos and Culture"
Steven J. Dick
Chief Historian, NASA (Retired)

Video: https://pswscience.org/meeting/cosmos-and-culture/

2011 – Meeting #2,286 – May 20, 2011
"2020 Vision: A Decadal View of New Worlds and New Horizons in Astronomy and Astrophysics"
Debra Elmegreen
Maria Mitchell Professor of Astronomy, Vassar College

Video: https://pswscience.org/meeting/2020-vision/

2012 – Meeting #2,302 – May 25, 2012
"Synthesizing Life"
Lee Cronin
Gardiner Professor of Chemistry, University of Glasgow

Video and Minutes: https://pswscience.org/meeting/synthesizing-life/

2013 – Meeting #2,318 – May 10, 2013
"Curiosity's Robotic Exploration of Mars' Gale Crater"
John P. Grotzinger
Fletcher Jones Professor of Geology
Chief Mission Scientist, Mars Science Laboratory, aka "Curiosity"

Video: https://pswscience.org/meeting/curiositys-robotic-exploration-of-mars-gale-crater/

2014 – Meeting #2,334 – May 9, 2014
"A Worm's Tale: Secrets of Evolution and Immortality"
Craig C. Mello
Blais University Chair in Molecular Medicine, University of Massachusetts Medical School
2006 Nobel Prize in Medicine or Physiology

Video: https://pswscience.org/meeting/a-worms-tale-secrets-of-evolution-and-immortality/

2015 – Meeting #2,348 – May 8, 2015
"Synthetic Biology: The Technology Engine of the 21st Century"
Pamela Silver
Professor, Systems Biology, Harvard Medical School

Video and Minutes: https://pswscience.org/meeting/synthetic-biology/

2016 – Meeting #2,362 – April 15, 2016
"The Exploration of Pluto"
Alan Stern
Principal Scientist, New Horizons
Vice President for Research, Southwest Research Institute

Video and Minutes: https://pswscience.org/meeting/the-exploration-of-pluto/

2017 – Meeting #2,379 – May 12, 2017
"Exoplanets and the Search for Life: An Exploration in Four Lectures"

Mark Clampin
Deputy Director, Sciences and Exploration Directorate
NASA Goddard Space Flight Center

Debra Fischer
Professor of Astronomy, Yale University

Hannah Wakeford
NASA Postdoctoral Fellow, NASA Goddard Space Flight Center

Shawn Domagal Goldman
Research Space Scientist, NASA Goddard Space Flight Center

Aki Roberge
Research Astrophysicist, NASA Goddard Space Flight Center

Video and Minutes: https://pswscience.org/meeting/exoplanets-and-the-search-for-life/

2018 – Meeting #2,394 – June 15, 2018
"Cassini's Spectacular Saturn How it Revolutionized Our View of Planets, Moons, and the Possibilities for Life in the University"

Julie Webster
Manager, Cassini Spacecraft Operations Office
Chief Engineer, Juno Mission to Jupiter, Jet Propulsion Laboratory

Phillip Nicholson
Professor of Astronomy
Acting Director of the Center for Astrophysics & Planetary Science, Cornell University

Bonnie Buratti
Principal Scientist and Manager, Planetary Science Section, Jet Propulsion Laboratory

Terry Hurford
Planetary Scientist
Geology, Geophysics and Geochemistry Laboratory, NASA Goddard Space Flight Center

Conor Nixon
Planetary Scientist, NASA Goddard Space Flight Center

Video and Minutes: https://pswscience.org/meeting/cassinis-spectacular-saturn/

2019 – Meeting #2,410 – May 17, 2019
"The Big Bang and Inflation: Glimpsing the Beginning of time from the Ends of the Earth
Brian Keating"
Professor of Physics, Center for Astrophysics & Space Sciences, University of California at San Diego

Video and Minutes: https://pswscience.org/meeting/beginning-of-all-things/

2020 – Meeting #2,421 – May 15, 2020
"The Event Horizon Telescope: First Ever Images of a Black Hole"
Sheperd S. Doeleman
Founding Director, Event Horizon Telescope, Harvard-Smithsonian Center for Astrophysics

Video and Minutes: https://pswscience.org/meeting/2421/

2021 – Meeting #2,442 – June 4, 2021
"Quantum Gravity"
Carlo Rovelli
Director, The Quantum Gravity Group, Centre de Physique Théorique de Luminy, Aix-Marseille University

Video and Minutes: https://pswscience.org/meeting/2442/

2022 – Meeting #2459 – May 20, 2022
"Ingenuity: The First Flying Machine on Mars, First Controlled Powered Flight on Another Planet"
J. (Bob) Balaram
Chief Engineer of Ingenuity, JPL/CalTech

Video and Minutes: https://pswscience.org/meeting/2459/

2023 – Meeting #2477 – May 19, 2023
"Architecture for the New Space Age: Learning to Work and Live Beyond Earth"

Paul D. Kessler
Aerospace Technical Manager,
NASA

Eleanor E. Morgan
Space Architect & Program Manager, Lockheed Martin

Daniel Inocente
Senior Space Architect, Blue Origin

Raymond "Corky" Clinton
Senior Technical Advisor, Science & Technology Office, NASA Marshall Space Flight Center

Video and Minutes: https://pswscience.org/meeting/2477/

2024 – Meeting #2,496 – May 17, 2024
"Unlost – Recovering the Text of Burnt and Carbonized Scrolls: Reading Herculaneum Papyri using X-ray CT and AI"
Brent Seales
Professor of Computer Science, University of Kentucky

Video and Minutes: https://pswscience.org/meeting/2496/

PSW Lectures by Nobel Laureates

Hans Bethe
1967, Physics, Stellar Nucelosynthesis
PSW Meeting #1,378, April 24, 1953

Nicolaas Bloembergen
1981, Physics, Laser Spectroscopy and Non-Linear Optics
PSW Meeting 1975

Baruch Blumberg
1976, Physiology or Medicine, Hepatitis B Virus and Vaccine
PSW Meeting #2,163, May 16, 2003

Gerald M. Edelman
1972, Physiology or Medicine, Structure of Antibodies
PSW Meeting #1,997, May 15, 1992

Val L. Fitch
1964, Physics, CP Violation
PSW Meeting 1981

James Franck
1925, Physics, Experimental Verification of Atomic Structure by Electron Bombardment
PSW Meeting #1,115, March 13, 1937

Joachim Franck
2017, Chemistry, Single-Particle Cryo-Electron Microscopy
PSW Meeting #2,420, February 21, 2020

D. Carleton Gajdusek
1976, Physiology or Medicine, Etiology of Kuru via Cannibalism
PSW Meeting May 18, 1990

Walter Gilbert
1980, Chemistry, DNA Sequencing Methods
PSW Meeting #1,931, May 20, 1988

Tsung-Dao Lee
1957, Physics, Parity Violation in Weak Interactions
PSW Meeting 1960

Willard F. Libby
1960, Chemistry, Radiocarbon Dating
PSW Meeting #1,408, March 25, 1955

John Mather
2006, Physics, COBE and Measurement of Cosmic Background Radiation
PSW Meeting #2,223, May 11, 2007
PSW Meeting #2,267, March 19, 2010

Craig C. Mello
2006, Physiology or Medicine, RNA Interference
PSW Meeting #2,334, May 9, 2014

Marshall Nirenberg
1968, Physiology or Medicine, Deciphering the Genetic Code
PSW Meeting May 7, 1982

Williams Phillips
1997, Physics, Laser Cooling
PSW Meeting #1,970 November 30, 1970
PSW Meeting #2,193, May 13, 2005

George Porter
1967, Chemistry, Flash Photolysis Laser Spectroscopy
PSW Meeting 1967

Ilya Prigogine
1997, Chemistry, Non-Equilibrium Thermodynamics
PSW Meeting April 28, 1978

Isidor I. Rabi
1944, Physics, Nuclear Magnetic Resonance
PSW Meeting 1977

Adam Riess
2011, Physics, Discovery of the Accelerating Expansion of the Universe
PSW Meeting #2,261, December 4, 2009

William Shockley
1956, Physics, Development of the Transistor
PSW Meeting #1,332, May 13, 1950

Frazer Stoddard
2016, Chemistry, Topological Bound and the Synthesis of Molecular Machines
PSW Meeting #2,294, January 6, 2012

George P. Thomson
1937, Physics, Electron Diffraction
PSW Meeting 1956

Harold C. Urey
1934, Chemistry, Discovery of Deuterium
PSW Meeting #1,656, May 1, 1970

Eugene P. Wigner
1963, Physics, Theory of the Atomic Nucleus
PSW Meeting 1968

FROM THE TELEPHONE PATENT RACE TO QUANTUM COMPUTING

Elish Gray Demonstrates his Telaphon
100th Meeting
February 12, 1876

"Mr. Elisha Gray gave an exhibition of a telaphon, which he had invented, by which musical tones transmitted over a telegraph wire can be responded to at a distant point."

110th Meeting. 1877. *Bulletin of the Philosophical Society of Washington* (2): 66-67.

Gray's demonstration was the first public demonstration of the telephone. On the Monday (February 14) following that Saturday (February 12) demonstration at the Society, Gray's lawyer went to the Patent Office to file a caveat for a telephone using a liquid transmitter. Alexander Graham Bell's lawyer also filed, later that day. Who was entitled to the patent, eventually granted to Bell, is still disputed today.

Alexander Graham Bell Demonstrates his Telephone
116th Meeting
January 13, 1877

"Mr. Alexander G. Bell, of Boston, made a communication on the telephone, which he had invented, exhibiting and describing its construction and explaining the principles on which it is operated."

116th Meeting. 1877. *Bulletin of the Philosophical Society of Washington* (2): 102-104.

"Quantum Information Science and Quantum Computing: The National Initiative"
Charles Tahan

Assistant Director for Quantum Information Science
White House Office of Science and Technology Policy

2,460th Meeting
June 3, 2022

100TH MEETING. FEBRUARY 12, 1876.

The President in the Chair.

Thirty-eight members and visitors present.

Mr. E. K. GILBERT made a communication on

THE HORARY OSCILLATIONS OF THE TEMPERATURE OF THE ATMOSPHERE,

in which he correlated barometric and thermometric changes and rates of change in observations made at San Francisco and Philadelphia in the equinoctial months of March and September

Mr. F. F. JUDD made a communication on

THE WATERSHED OF THE ADIRONDACK REGION,

giving results of a survey of the head-waters of the Hudson and Rackett Rivers by Prof. FARRAND N. BENEDICT in 1874, and dwelling particularly on the feasibility and facility of converting the numerous and extensive lakes in that region into reservoirs, so managed as to greatly reduce freshets in the Hudson in the spring, and supply the deficiency of water in the fall.

Remarks were made by Messrs. HARKNESS, HENRY, POWELL, and ALVORD.

Mr. ELISHA GRAY gave an

EXHIBITION OF A TELAPHON,

which he had invented, by which musical tones transmitted over a telegraph wire can be responded to at a distant point. Several tones of a different pitch were transmitted simultaneously by a single circuit from one room, and each responded to in another room by a reed attached to an electric magnet and tuned to the same pitch. With such apparatus several messages can be sent at the same time by the same circuit, and at the place of reception each would be distinguished from the rest by difference of tone and by being responded to by a magnet whose reed was tuned to the same pitch as that from which it had come. He stated that messages had been thus transmitted successfully by a circuit of 300 miles.

He also exhibited other methods by which the sounds could be responded to; in one of which a tune played in one room was repeated by a diaphragm in another, with which an electrical connection was made.

116TH MEETING. JANUARY 13, 1877.

The President in the Chair.

Forty-nine members and visitors present.

Mr. G. K. GILBERT made a communication on

LAKE BONNEVILLE,

the great fossil lake of Utah. He described an ancient outlet of the lake at Red Rock Pass near the town of Oxford, Idaho, by which its waters were discharged into Snake River. During and since the desiccation of the lake, the land which it covered has been tilted to the northward, in common with the region of the Laurentian lakes and the eastern and western seaboards. He further described a small movement along the line of the great fault at the western base of the Wasatch Mountains, by which the altitude of the mountains above the adjacent valley has been increased at a date far more recent than that of the ancient lake. (*A full account of his observations will appear in the publications of the U. S. Geographical and Geological Survey of the Rocky Mountain Region, in charge of Prof.* J. W. POWELL.)

Remarks were made by Mr. ANTISELL on the channel described by Mr. GILBERT as affording temporary drainage; and by Mr. ALVORD on the appropriateness of giving to this basin the name of BONNEVILLE, who was the first to make a scientific exploration of this region. Great Salt Lake for many years appeared on the maps as Lake Bonneville.

Mr. ALEXANDER G. BELL, of Boston, made a communication on

THE TELEPHONE,

which he had invented, exhibiting and describing its construction and explaining the principles on which it is operated. The sound of the human voice received on a small plate of thin Russian sheet-iron was conveyed by a telegraphic wire to a similar apparatus in another room and repeated by the vibrations of a similar iron plate. He stated that he made use of an undulatory, instead of an intermittent current, that no battery was necessary, but that the variations of intensity were produced by the vibrations of the soft iron plate, varying its distance from the poles of an electro-magnetic helix just behind it. He stated that the experiment had been successfully conducted, where the operator and hearer were 153 miles apart.

He referred also to an experiment made by Prof. HENRY many years ago, where an air played on one piano was repeated by another on the opposite side of the street, a rod of soft pine in contact with the sounding-boards of each forming the connection.

Mr. HILGARD and Mr. HENRY spoke of the value and astonishing character of Mr. Bell's discovery and invention.

Mr. BELL having spoken of the difficulty with such consonants as *p*, *t*, and *k*, Mr. MASON suggested that such an instrument, when perfected, might be used in analyzing linguistic sounds.

"Quantum Information Science and Quantum Computing: The National Initiative"

Charles Tahan
Assistant Director for Quantum Information Science
White House Office of Science and Technology Policy

2,460th Meeting
June 3, 2022
https://pswscience.org/meeting/2460/

Minutes of the 2,460th Meeting

On June 3, 2022, in the Powell Auditorium at the Cosmos Club in Washington, D.C., President Larry Millstein called the 2,460th meeting of the Society to order at 8:03 p.m. EDT. He welcomed new members, and the recording secretary read the minutes of the previous meeting.

President Millstein then introduced the speaker for the evening, Charles Tahan, Assistant Director for Quantum Information Science and Director of the National Quantum Coordination Office, White House Office of Science and Technology Policy. His lecture was titled, "Quantum Information Science and Quantum Computing: The National Initiative."

Tahan said quantum mechanics enabled the physical construction of modern computing and communication. But, until recently, computers operated on systems incapable of efficiently simulating quantum mechanics. In recent years, quantum mechanics research has evolved into information science.

Tahan said quantum information science is a foundational shift in information technology, which will have unpredictable impacts on myriad fields. For example, he said, quantum computers will factor very large numbers and render current encryption technologies useless. Building such powerful computers will require a significant investment in technology-related fields, such as quantum physics, microelectronics, and electrical engineering.

The speaker then explained the basic science of quantum computing. A quantum bit, or qubit, is a two-level atom in its lowest state which can quantum mechanically be in the superposition of both zero and one at the same time. But when measured, a qubit is read only as zero or one at any time. As qubits are added together, their system size grows exponentially. For that reason, quantum computing that relies on qubits is enormously more powerful than classical computing that relies on regular bits.

Tahan then described the many different potential realizations for constructing an actual quantum computer. He focused on superconducting circuits and electrons, two potential realizations that he said are analogous to modern-day fabrication of computer chips. Tahan said current quantum computing error rates are 10 to 12 orders of magnitude away from being able to perform "useful" calculations.

Tahan focuses his work on solid state qubits. He prefers spin qubits, which are the most analogous to transistors in modern computers. They are simple, move quickly, and are compact. Within the spin qubit category are qubits with different spins. Certain multiple qubits may also be combined to form molecular qubits, which are extremely powerful but difficult to read. Tahan then described the modulated readout approach his team developed with collaborators from the University of Wisconsin.

Current large technology companies like Amazon and Google are using superconducting qubits to build quantum computers. Tahan described these qubits as "just wires on a chip," which are now as small as 20 nanometers across and superconduct at extremely low temperatures. He then described Google's recent quantum computing accomplishments, and expressed optimism for future research into combining the best features of superconducting and semiconducting qubits.

In response to the National Quantum Initiative, the Laboratory for Physical Sciences (LPS) established the LPS Qubit Collaboratory, to pursue disruptive fundamental research, build a quantum workforce, and to grow collaborative partnerships to tackle the difficult problems in quantum research. Rather than ask for proposals to then fund, the

Collaboratory presents problems and invites collaborators to join in the work.

One of the Collaboratory's principal goals is to develop quantum qubit chips in the same way standard chips are currently manufactured. Tahan described how the Collaboratory reduced the problem into an approachable one, and the advances so far.

The National Quantum Initiative Act was signed into law in 2018, and created 13 new centers across the country and established a committee to coordinate multiple bodies in pursuing quantum research and development. Tahan said the three principal aims of the current U.S. Administration are to get the science right, enhance American competitiveness, and build a workforce to develop a new quantum field of work. He then described current policy efforts to achieve those ends.

The speaker then answered questions from the in-person and online viewing audience. One member asked whether software or hardware were the immediate obstacles to further quantum research. The speaker said hardware is currently the biggest challenge, and that he believes one of the approaches he presented on will work.

A guest asked why the speaker's team used germanium instead of materials more commonly used in superconductors. The speaker said that holes in germanium's structure avoid the indirect band-gap issue presented in silicon quantum computing.

After the question and answer period, President Millstein thanked the speaker, made the usual housekeeping announcements, and invited guests to join the Society. President Millstein adjourned the meeting at 10:02 p.m.

Temperature in Washington, D.C.: 24.5° C
Weather: Mostly Sunny
Number in-person and live online viewers: 109, and views on the PSW Science YouTube and Vimeo channels in the first two weeks: 433.

Respectfully submitted,
James Heelan, Recording Secretary

THE FIGHT AGAINST MALARIA

King Announces the Vector for Malaria

"The Prevention of Malarial Diseases, Illustrating, inter alia, The Conservative Function of Ague"

Albert F.A. King
National Medical College
Columbian University (now George Washington University)

229th Meeting
February 10, 1883

"That material diseases were ever produced solely by the *inhalation* of supposed poisonous vapors was held to be untenable. . . . The proboscis of the mosquito was Nature's inoculating needle." (9)

229th Meeting. 1884. *Bulletin of the Philosophical Society of Washington* (6): 5-10.

James Tackles Malaria with Synthetic Biology and CRISPR

"The Mosquito, Synthetic Biology, CRISPR, and Malaria: Using Gene Modification and Gene Drive to Eradicate a Disease"

Anthony A. James
Distinguished Professor
University of California - Irvine

2,378th Meeting
April 28, 2017
https://pswscience.org/meeting/the-mosquito-synthetic-biology-crispr-and-malaria-2/

229TH MEETING. FEBRUARY 10, 1883.

The President in the Chair.

Forty-two members and visitors present.

It was announced that reports of the scientific proceedings would hereafter be furnished to Science.

Mr. W. H. DALL announced that an opportunity would be afforded members to contribute to the Balfour Memorial Fund.

A communication was then read by Mr. A. F. A. KING on

THE PREVENTION OF MALARIAL DISEASES, ILLUSTRATING, *inter alia*, THE CONSERVATIVE FUNCTION OF AGUE.

[Abstract.]

The various theories thus far presented in explanation of the phenomena of malaria were unsatisfactory and insusceptible of scientific demonstration.

According to the best medical authorities the most generally admitted facts upon which the present orthodox theory of malaria rests were as follows: 1. Malaria affects by preference low and moist localities. 2. It is almost never developed at a lower temperature than 60° F. 3. Its evolution or active agency is checked by a temperature of 32° F. 4. It is most abundant and most virulent as we approach the equator and the sea-coast. 5. It has an affinity for dense foliage, which has the power of accumulating it, when lying in the course of winds blowing from malarious localities. 6. Forests or even woods have the power of obstructing and preventing its transmission under these circumstances. 7. By atmospheric currents it is capable of being transported to considerable distances—probably as far as five miles. 8. It may be developed in previously healthy places by turning up of the soil, as in making excavations for the foundations of houses, tracks for railroads, and beds for canals. 9. In certain countries it seems to be attracted and absorbed by bodies of water lying in the course of such winds as waft it from the miasmatic source. 10. Experience alone can enable us to decide as to the presence or absence of malaria in any given locality. 11. In proportion as countries, previously malarious, are cleared up and thickly settled, periodical fevers disappear, in many instances to be replaced by typhoid fever (?) 12. Malaria usually keeps near the surface of the earth. It is said to "hug the ground," or "love the ground." 13. It is most dangerous when the sun is down, and seems almost inert during the day. 14. The danger of exposure after sunset is greatly increased by the person exposed *sleeping* in the night air. 15. Of all human races the white is most sensitive to marsh fevers, the black least so. 16. In malarial districts the use of fire, both indoors and to those who sleep out, affords a comparative security against malarial disease. 17. The air of cities in some way renders the poison innocuous; for, though a malarial disease may be raging outside, it does not penetrate far into their interior. 18. Malarial diseases are most prevalent towards the latter part of summer and in the autumn. 19. Malaria is arrested not only by trees, but also by walls, fences, hills, rows of houses, canvas curtains, gauze veils, mosquito nets, and probably by fishing nets. 20. Malaria spares no age, but it affects infants much less frequently than adults.

These generally admitted facts were insusceptible of scientific explanation by the marsh fever hypothesis of Lanscisci; but were capable of explanation by the theory that marsh fevers are produced by the bites of proboscidian insects, notably in this and in some other countries by mosquito bites.

A review of the natural history, habits, and geographical distribution of the mosquito was next presented in explanation of the twenty statements above quoted.

In discussing statement 15, it was maintained that the comparative immunity of the black races was largely due to color, the dark complexion of the skin being another illustrative instance of "protective coloring" so often observed in other animals, and by which, in this instance, the negro was protected from the *sight*, and consequently from the *bite* of the mosquito; a similar protection being further secured by the offensive odor and greasiness of his cutaneous secretions, aided by artificial inunction of the body with grease, paint, pitch, &c., which last probably constituted the initial step in the evolution of dress. Hence malarial melanosis was considered to be the designed natural termination of ague—its conservative function—destined to modify the individual by defensive adaptation against the mosquito, whose penetrating proboscis, like an inoculating needle, infected the body with malarial poison, no matter whether this last was mosquital saliva, the *Bacillus malariæ* of Klebs and Crudelli, or some other element as yet unknown.

The spleen, whose function is not yet settled by physiologists, was regarded as the chief *pigment-forming organ*, and was *designed for this purpose in the economy of the organism.* Generally considered a superfluous organ, capable of removal without any great interference with the functions of the organism, it was naturally designed to *meet the emergency of variation in skin-color* to secure "protective coloring" against fever-producing proboscidian insects, as before indicated. The natural process, however, required exposure of the naked body to the sun during the chill stage, in order to secure deposit of the newly formed pigment in the skin. Nature had not anticipated the artificial appendage of dress, and the organism had not inherited from ancestral progenitors any provision for so unexpected an addition. Chills do not occur at night, but only between the rise and setting of the sun; sunlight during the chill stage being a necessary requirement, in order that nature's design of cutaneous chromatogenesis may be consummated. Other racial differences between the whites and blacks—such as even cerebral capacity and variations in the skeleton—might be susceptible of explanation by blood changes resulting from malaria. The marrow of bones was also a pigment-forming tissue, and the aching of bones during ague, especially in so-called "break-bone" fever, suggested congestion and modified nutrition in the osseous structures, such as might eventually lead to modification in the skeleton. The inhabitants of oriental countries especially were more vigorous and intelligent if they lived in elevated regions, than were others inhabiting mosquito-infected lowlands and sea coasts.

In further support of the mosquital origin of malarial fevers numerous noted medical authorities were cited, showing that, in all parts of the world where these diseases prevail, immunity was secured by protecting the body from mosquito bites. The geographical distribution and seasonal evolution of mosquitoes and other proboscidian insects were shown partially to agree with the times and places in which malarial diseases prevail; though from lack of information conclusive evidence on this point was yet wanting. There was, however, a general admission on the part of medical authorities that swarms of these insects in almost any locality were a pretty sure sign of malignancy.

On the other hand numerous instances were adduced from "Narratives" and "Travels" in which the bodies of persons had been covered with pustules, "resembling small-pox," from mosquito bites without any subsequent occurrence of fever having been *recorded* by the narrating authors.

This opposing evidence was inconclusive, (1) because the authors cited were not in search of medical information; (2) because the period of incubation, being often long and uncertain, fever *may* have occurred after the mosquito bites had been forgotten; (3) the insect proboscis (like a vaccine lancet unarmed with virus) might be uncontaminated with fever poison, or fever germs; and (4) *successful* inoculations of specific germ poisons are *not* usually followed by immediate local suppuration at the point of puncture, but only after a certain period of incubation, the *immediate* local inflammation being rather *preventive* of subsequent blood infection.

The possible spread of yellow-fever contagion by the inoculating proboscis of the mosquito carrying infecting matter drawn from the blood of yellow-fever patients to unaffected persons was suggested. In epidemics, the spread of the disease stopped as soon as a freezing temperature paralyzed the mosquito, &c.

The spread of spotted-fever, typhus-fever, in jails, ships, &c., was referred to the inoculating instrument of fleas, &c.—these insects usually prevailing among filthy people thickly crowded together.

That malarial diseases were ever produced solely by the *inhalation* of supposed poisonous vapors was held to be untenable. Experimenters, who had demonstrated the existence of specific poisons for special fevers, had equally proven that the mode by which such poisons, when obtained, could be introduced into the body for the artificial production of disease, was by *inoculation through the skin.* These experiments were imitations of insect inoculation. The proboscis of the mosquito was Nature's inoculating needle.

The *modus operandi* of the eucalyptus tree in preventing malarial diseases was ascribed tentatively to the tree being destructive to, or interfering directly or indirectly with, the propagation and development of mosquitoes.

From the foregoing conceptions as to the origin of malarial disease, the following prophylactic measures were deducible:

1st. *Personal* protection from all winged insects, especially during evening and night, by gauze curtains, veils, window-blinds, or clothing impenetrable by the proboscis of inoculating insects; and further, personal protection both from these and all creeping insects, especially during epidemics, endemics, and in crowded jails, ships, &c., by daily inunction of the whole body with some terebinthinate, camphorated, or eucalyptalized ointment or liniment.

2d. *Domiciliary* protection (*a*) *exteriorly*, by screens of trees, walls, fences, &c., interposed at some distance between dwellings and the supposed sources of malaria, or mosquito nurseries; and with fires or lamps arranged as traps for the attraction and destruction of such winged insects as may encroach nearer. A further protection (*b*) in the *interior* of dwellings being secured by the use of smoke (as of tobacco or prethrum) or of some volatile aromatic substance, as of camphor, assafœtida, garlic, &c., which may be offensive to proboscidian intruders.

3. *Municipal* protection by groves of trees (pines, cedars, or eucalyptus) planted between cities and the sources of malaria and mosquitoes, together with cordons of electric or other lights, between said grove and the marsh, the lights to be arranged as fly-traps for the retention and destruction of such winged insects as may be thus secured.

With relation to the city of Washington, it was suggested that the Washington monument would afford a good opportunity (by placing illuminated fly-traps at different elevations on its exterior) for ascertaining the height at which mosquitoes fly, or are brought by the wind from the adjacent Potomac flats. The proposed reclamation of the flats could scarcely do more than mitigate malarial disease, so long as our summer and autumn southern breezes come, laden with mosquitoes, from the miles of unreclaimed swamps farther down the river, as at Four-mile Run and other nearer localities.

Mr. BILLINGS remarked that, since ague did not invariably result from insect bites, the most that could be claimed was that they accomplished an accidental inoculation with malarial poison.

The subject was also discussed by Messrs. DOOLITTLE, TONER, and ANTISELL.

The meeting closed with an exhibition by Mr. C. E. DUTTON of a series of oil paintings illustrative of the Hawaiian Islands.

"The Mosquito, Synthetic Biology, CRISPR, and Malaria: Using Gene Modification and Gene Drive to Eradicate a Disease"

Anthony A. James
Distinguished Professor
University of California - Irvine

2,378th Meeting
April 28, 2017

Minutes of the 2,378th Meeting

President Larry Millstein called the 2,378th meeting of the Society to order at 8:11 p.m. He announced the order of business and welcomed new members. President Millstein presented a summary of the 37th meeting of the Society, held in 1872. The minutes of the previous meeting were read and approved. President Millstein then introduced the speaker for the evening, Anthony A. James, a Distinguished Professor in the Departments of Microbiology & Molecular Genetics and Molecular Biology & Biochemistry at the University of California, Irvine. His lecture was titled "The Mosquito, Synthetic Biology, CRISPR, and Malaria".

James began by explaining that malaria has afflicted humans for all of recorded history, with references to deadly recurrent fevers appearing in Mesopotamian cuneiform tablets from 6000 BC. The famous Greek physician Hippocrates noted the link between swampy waters and fever cycles.

Malaria is a bloodborne infectious disease resulting from a protozoan parasite which is transmitted by mosquitos. Today there are five species of malaria parasite that affect humans, the newest of which crossed over to humans within the last 15 years. The two species most deadly to humans are plasmodium falciparum and plasmodium vivax. Importantly, there are no free-living forms of the parasite; it exists only inside a suitable host. Malaria is only spread by adult female mosquitos, which feed on blood as a source of protein for nutrition and egg production.

Malaria today is endemic in the poorest countries of the world, existing in all tropical and subtropical regions. In 2015, there were 212 million cases worldwide, resulting in approximately 429,000 deaths, 90% of which were in sub-Saharan Africa. Although we have many existing tools that are effective against malaria—including bed nets, spraying, eliminating standing water, and anti-malarial drugs—for various practical, economic, and political reasons, these tools have been insufficient to control malaria in many countries. To date, no effective vaccines have been developed against any protozoan parasite, including malaria.

James then described a promising new strategy of engineering mosquitos with genes that prevent transmission of the malaria parasite. To do this, scientists first searched for mosquito genes that are activated when the mosquito feeds; they found suitable genes related to digestion. Next, they identified a gene in mice that reacted very strongly to human strains of the malaria parasite, neutralizing it. Splicing this mouse gene into a mosquito creates a transgenic mosquito that express antibodies in the midgut and the salivary gland every time the mosquito feeds, curing the mosquito and eliminating transmission back to humans.

After developing a malaria-immune mosquito, the next task is to distribute these genes throughout the mosquito population. This can be done simply by breeding and releasing millions of mosquitos, but a more effective strategy is to include a "gene drive" in the engineered mosquito's DNA that ensures that the anti-malarial genes are preferentially inherited when the mosquito breeds, overriding the traditional Mendelian pattern of inheritance.

The creation of a gene drive is possible because of advances in the use of the CRISPR/Cas9 gene editing system, which consists of the Cas9 nuclease, which cuts DNA, plus guide RNA, which targets the cuts to a precise position in the mosquito DNA sequence. When the cell attempts to repair the cut DNA, the most available substrate for reconstituting the cut section is the lab-created plasmid containing the anti-malarial mouse gene. Moreover, this engineered sequence also encodes the cas9 nuclease and guide RNA, which means that the repaired DNA will copy itself autonomously to the other, unmodified chromosome, creating mosquitos that have a one hundred percent chance of passing on the engineered genes.

James explained that in addition to the functional parts of the engineered plasmid—the nuclease, guide RNA, and malarial resistance sequence—it also contains a protein for red coloring from a jellyfish that allows engineered mosquitos to be readily distinguished from wild-type mosquitos.

James noted that drugs, insecticides, and vaccines all have well-established regulatory pipelines from discovery to development to delivery, but genetics based vector control does not. Nonetheless, James expressed confidence that carefully constructed, rigorously conducted phased trials will demonstrate to regulators, governments, and communities that genetically-engineered mosquitos are a safe and effective malaria control strategy that, when combined with existing tools, could ultimately enable world health authorities to finally eradicate malaria.

After the conclusion of the lecture, President Millstein invited questions from the audience.

One questioner asked how the mouse gene confers malaria resistance in the mosquito. James explained that the engineering confers two forms of resistance. First, it creates antibodies that bind to surface proteins that the parasite uses to invade cells, thereby denying it entry into the host cells. The second method simply coats the surface of the parasite in antibodies, preventing it from interacting with the host cells.

Another questioner as why the engineered mosquitos used genes for mouse antibodies and not human antibodies. James explained that human resistance acts only during the human blood cell stages and does not provide complete protection, so the mouse gene was chosen to target malaria completely and at the point of transmission.

After the question and answer period, President Millstein thanked the speaker, made the usual housekeeping announcements, and invited guests to join the Society. At 9:59 p.m., President Millstein adjourned the 2378th meeting of the Society to the social hour.

Attendance: 105
The weather: Partly cloudy
The temperature: 24°C

Respectfully submitted,
Preston Thomas
Recording Secretary

NEUTROINOS AND OTHER PARTICLES

"Mesons and Nuclear Forces"

Hans Bethe
Cornell University
1967 Nobel Prize in Physics

1,378th Meeting
April 24, 1953

Bethe, Hans. 1954. "Mesons and Nuclear Forces." *Journal of the Washington Academy of Sciences* 44 (4): 97-105.

"Ice-Cube – The Under-the-Ice Neutrino Observatory in Antarctica: Opening a New Window on the Universe from the South Pole"

Frances Halzen
Professor of Physics
University of Wisconsin-Madison

2,466th Meeting
November 4, 2022
https://pswscience.org/meeting/2466/

JOURNAL

OF THE

WASHINGTON ACADEMY OF SCIENCES

Vol. 44 April 1954 No. 4

PHYSICS.—*Mesons and nuclear forces.*[1] HANS A. BETHE, Cornell University. (Communicated by Richard K. Cook.)

1. HISTORY

I want to start out with some brief remarks about the history of the subject of mesons and nuclear forces. It is an example, as good as any I know in recent scientific progress, of both the wisdom and the folly of scientists. The theory of nuclear forces began in 1932 with the discovery of the neutron. This made possible a consistent picture of the structure of the nucleus, namely, to consider the nucleus as composed of neutrons and protons which are held together by very strong forces, different from and stronger than any other forces which we had known in nature before. Only three years after the discovery of the neutron and the start of nuclear theory, Yukawa suggested that the nuclear forces were transmitted between the nuclear particles, the neutron and the proton, by other particles, as yet undiscovered, which have now come to be known by the name of mesons. Yukawa predicted that there should be such particles, that they should have a mass of 100 to 200 times the electron mass, that they should be charged, and that they should have integral spin, probably either zero or one.

Three years later, Yukawa's prediction came true. Particles were discovered in cosmic radiation by two groups of people, Anderson and Neddermeyer, working at the California Institute of Technology, and Street and Stevenson, working at Harvard University. These particles had a mass of about 200 electron masses, they had a positive or a negative charge just as Yukawa had wanted, and they seemed to fulfill pretty well his program. In the succeeding nine years experimental physicists kept discovering more and more properties of these particles and theoretical physicists kept calculating what such particles would do for nuclear forces. Only the twain never met; the predictions of the theorists were completely different from the way the particles actually behaved.

The theorists predicted that these particles, the mesons, should interact very strongly with nucleons once they were formed and should, therefore, be easily absorbed, be easily scattered, and should easily cause nuclear reactions. They did nothing of the kind. In fact, they did nothing of any kind. They just moved along, were slowed down as any charged particle is, finally came to a stop and disintegrated in some way then unknown. All the same, theoretical physicists persisted in their belief in the connection between Anderson's cosmic-ray particle, and Yukawa's prediction. But the differences between the experimental results and the theoretical prediction were so great that it was proposed to make the best of another disagreement among physicists, one about the name of the particle, and to call the experimental particle, the mesotron, and the theoretical particle, the meson. Attempts to identify the two particles continued until finally, in 1947, a group of Italian physicists, Conversi, Piccioni, and Pancini, found that the cosmic-ray meson had even less interaction with the nucleus than had been suspected before. Even when they gave the meson a chance of sitting around the nucleus for a long, long time, namely a microsecond, even then it wouldn't

[1] The Twenty-second Joseph Henry Lecture of the Philosophical Society of Washington, delivered before the Society on April 24, 1953. This Lecture has also appeared in Physics Today **7** (2). 1954.

make use of its opportunity and would not get captured by the nucleus. This finally proved that the two particles could not be identified. But then it took only a few months before the solution to the puzzle was found in Bristol, England, by Occhialini, Powell, and Lattes, who discovered that there existed still another particle which they called the π-meson, the primary meson, which decayed after a short time into the meson which had been previously discovered and which they called the μ-meson, the "meson" meson.

The π-meson, at last, fulfilled Yukawa's dream. It had a strong interaction with the nucleus. Once created, it was easily scattered; it was easily absorbed by the nucleus; and since 1947 we have come to believe that the π-meson really is the particle which transmits nuclear forces. Many properties of the π-meson were discovered by the Bristol group, working with cosmic ray mesons and photographic plates, but the greatest progress about finding out the properties of this particle was only made after this particle was produced artificially in accelerators, first at Berkeley, then also at other laboratories. This again was an example of international cooperation: a powerful synchro-cyclotron, was available in Berkeley, but the Berkeley physicists did not discover artificially produced π-mesons until the special technique of reading photographic plates was brought to them from England by a Brazilian, Lattes. Since that time we have learned quite a lot about π-mesons, and this is what I want to talk about. However, before the main subject, it will be good to review some of the properties which had been predicted for the π-meson by nuclear physics on the basis of pure theory.

2. PREDICTIONS FROM NUCLEAR PHYSICS

In the first place, from nuclear physics it was predicted that the meson transmitting nuclear forces should exist in three forms, positively charged, negatively charged, and neutral. The positive and negative variety are easily visible. The neutral variety was only a theoretical prediction until 1949, based upon a very fundamental property of nuclear forces, namely, the so-called charge-independence. In 1935, it was discovered in the Department of Terrestrial Magnetism of the Carnegie Institution of Washington that the forces between two protons are just about the same as the forces between proton and neutron provided the pair of particles is in the same state of motion with respect to each other. This fact has, in the meantime, been established by many other pieces of experimental evidence, and by the theoretical work of Breit and others, and is known as the theorem of charge independence. Now if there were only charged mesons then there could only be processes of the type that Yukawa had predicted, namely, the following: A proton can emit a positive meson, thereby turning into a neutron, and then positive meson can then be absorbed by some neutron which may be in the neighborhood and which thereby is changed into a proton. A second *proton* in the neighborhood could not absorb the positive meson because thereby it would acquire two charges and that would give a particle which presumably does not exist. So in this way we can have an interaction between a proton and a neutron, but not between two protons. When proton and neutron interact, they exchange their charge, a kind of interaction which had been postulated in nuclear physics even before Yukawa and was known as exchange force. In fact the exchange character of the forces was one of the clues which Yukawa had when he invented his theory. However, you can see that by the exchange of charged mesons you cannot, in first order, get any interaction between two protons or between two neutrons; you need a neutral particle in order to transmit such interaction. That a neutral meson should exist was first postulated by Kemmer in England who set up a theory known as the "symmetric meson" theory. In this theory, neutral and charged mesons of either charge are all presumed to be coupled to the nucleon, and the strength of the coupling is supposed to be the same for all. (The proton and the neutron are both called by the generic name "nucleon.") Kemmer's theory leads indeed to charge-independence of nuclear forces.

After physicists were able to produce mesons artificially they found in fact the neutral meson in addition to the charged mesons. The neutral meson unfortunately cannot be seen directly because it has an extremely short life. It decays within something like 10^{-15} seconds into two gamma rays. The short lifetime can be understood theoretically but I will not go into this problem.

A second point which nuclear theory predicted about mesons was that the mesons should not be scalar; a scalar particle is one which has no spin, in contrast to electrons and nucleons which have a spin of one-half—i.e., they have an angular momentum "around their own axis", somewhat like a spinning top. The term "scalar" implies more than the absence of spin; it also implies that the wave function of the particle remains unchanged when the entire space is reflected on a mirror plane. There is also the possibility that the wave function of a spin-zero particle changes sign upon such reflection; in this case, the particle is called "pseudoscalar." Now scalar mesons could be excluded because the nuclear forces resulting from them are central forces depending on the separation of the two nucleons and on nothing else, whereas experimentally the forces are found to depend also on the direction of the spins of the two nucleons. The third prediction of nuclear theory was that the mesons should be pseudoscalar. This was deduced, in fact, from the detailed dependence of nuclear forces on the direction of the nucleon spins, in particular from the sign of the quadrupole moment of the deuteron.

A fourth prediction comes from an entirely different field of physics—not from nuclear physics but from an investigation of the fundamental properties of fields which has been pursued with some success especially since 1947. The problem is the mathematical treatment of the interaction of two fields like the electric field, the field of mesons, of nucleons, of electrons, etc. Before 1947, the treatment of such interactions gave certain infinite results, but since 1947 we do know how to deal with these infinities. This is done by the so-called "theory of renormalization" in which the quantities which used to turn out infinite are re-interpreted as a change of the mass and the charge of the particle. After such re-interpretation, one can then show that all physically observable quantities are, in fact, finite. This modern theory of fields is further able to tell the sheep from the goats: certain theories *can* be made finite by renormalization, whereas others are intractable. You can write the fundamental equations of these other theories on paper, but when you try to calculate the probability of—let's say—the scattering of a meson by a nucleon, you will invariably find an infinite result. So this development of the theory of fields has given us a principle of selecting between possible and impossible expressions for fields and their interactions. In particular, a field consisting of charged particles, like mesons, which can be emitted and absorbed by other charged particles (nucleons), can only have spin zero. So in this respect the theory of fields confirms what nuclear physics had already postulated, namely, that mesons should have spin zero. And it said one more thing, about the coupling of the mesons with the nucleons, which I will discuss later.

3. PROPERTIES OF FREE MESONS

Now let us examine the experiments on mesons. There are many types of experiments that you can do with particles. The type of experiment which I shall discuss first is a simple one, which can be interpreted without elaborate calculations—just the qualitative result is sufficient to give an answer. First of all, the mass of the meson has been very accurately determined; it is 273 electron masses for the charged meson and 264 for the neutral meson. The difference may perhaps be due to an electric self-energy, i.e., due to the interaction of the electric charge of the charged meson with the electric field. Secondly, we know that the mesons have only a finite lifetime; the charged π-mesons live about 10^{-8} seconds. This is relatively long and makes it possible to observe them because, if you have a charged meson of this lifetime moving with a velocity close to that of light, then it will on the average go 10 feet before it decays, which is a sufficient distance to make observations. On the other hand, the neutral meson which lives for 10^{-15} seconds will travel only about 10^{-5} cm, which is not enough to observe its properties before decay. The third property, which is a much more important one for our consideration, is the spin of the particle. This has been determined experimentally for both the charged and the neutral meson, by methods which are both ingenious and quite different.

The determination for the charged meson is based upon the statistical principle of detailed balancing. One knows from statistical mechanics and from quantum mechanics, that when you have a process which can go in one direction and then a process which can go the opposite way, the probabilities of these two processes must bear a close relation to each other. Namely, if all the types of particles involved were present at a very high temperature, such that there are equally many particles in each quantum state, then processes in both directions must occur equally often to preserve equilibrium. Now, what does this principle of detailed balancing have to do with the spin of the meson? If you have a particle of spin zero and if this particle is in an external field then there is only one way for this particle to behave: it can't orient anything with regard to the external field, it has only one quantum state. On the other hand, a particle of spin one can have three different orientations of its spin with regard to the external field. Now, let us consider a process in which a spin-one meson is produced and a process in which it is absorbed. If it is produced, it can be produced with three different directions of the spin. If it is absorbed, we know that it starts out with one definite direction of spin; therefore the ratio of the probabilities of being produced and being absorbed will contain a factor three if the particle has spin one, a factor which is known as the "statistical weight". If the particle has spin zero, on the other hand, the factor three is replaced by one, so that the ratio of production probability to absorption probability is three times smaller. This principle was suggested by Marshak for the investigation of the meson spin and was used successfully by two groups of experimental physicists, one at the University of Rochester and one at Columbia University. Both came out with the result that the meson, indeed, has spin zero as had been predicted by both nuclear physics and the general theory of fields.

A little more difficult was the next step, namely, to find out whether the meson is scalar or pseudoscalar; and now I must explain in more detail what this means. You probably know that atomic systems are characterized by a certain quantity which is known as the *parity*. The parity tells you how the wave function of the system behaves when you change the sign of all the coordinates, that is, when you change x into $-x$, y into $-y$, z into $-z$, i.e., when you make what is known as an inversion. In the case of zero angular momentum, you can do something simpler instead, namely, take a plane through the center of the atom and make a reflection of the whole space on this plane as if the plane were a mirror. Now let us ask what the wave function does when you make an inversion. We know that there are atomic states whose wave function does not change; we call these "even." There are other states whose wave function changes sign upon inversion, those we call "odd" states. This is the property that we call the parity. For instance, in optical spectra, any allowed spectral line leads from an even state of the atom to an odd state, or *vice versa*. Similarly, one defines the parity of nuclei. For instance, the deuteron consists of a neutron and a proton with a wave function which is even. At the same time the deuteron has an angular momentum of one, which comes from the spin of the nucleons being parallel, so we have an even state of spin one. Now suppose the deuteron is made to absorb a positive meson, then it will change into two protons. These two protons obey the Pauli principle, which says that the wave function of the whole system must be antisymmetrical, and which, therefore, says that there must be a certain relation between the spin of the system and the parity. It says, among other things, that it isn't possible to have any state of spin one and even parity but only of spin one and odd parity. This conclusion from the general principle of quantum mechanics is somewhat complicated to derive so I won't bother to do it.

Now, the way that the parity of the meson was determined was exactly by an experiment in which mesons were absorbed by a deuteron. To start with, negative mesons were allowed to be captured in a Bohr orbit around the deuteron. When this happens, the meson will finally go into the innermost Bohr orbit which has zero orbital momentum and afterwards, if you wait long enough, it will be captured by the deuteron. Since it has negative charge, the meson will convert the deuteron into two neutrons which will then leave in opposite directions. Then one can ask whether this capture process is permitted by conservation laws. Now, as I said, the charged meson has no angular momentum to contribute; so you start out with an angular momentum of one and with a state of even parity of the deuteron. Then in the end you get two neutrons, and as I have just said about two protons, so also two neutrons cannot exist in a state of angular momentum one and even parity. Therefore, this process ought to be forbidden by the rule of parity conservation and angular momentum conservation—unless the meson itself contributes something to the parity. But actually experimentally, the capture of negative mesons by deuterons occurs with great eagerness, and it does give two neutrons. So we have to conclude that the meson contributes something to the parity, that it changes the parity of the system. This is precisely what is meant by a pseudoscalar particle; it is one which changes the parity of the nucleon system when it gets absorbed or emitted by the system.

4. COUPLING OF MESONS AND NUCLEONS

Therefore, up to this point, the experiments had confirmed exactly what the theory of nuclear forces had predicted. Now the next point is a considerably more difficult one and is perhaps the most important question in meson theory. This is the question of *how* the mesons are coupled to the nucleon. The coupling of mesons—the coupling of any two fields—is expressed by a term in the Hamiltonian of the system, and I am afraid I have to get a little bit technical at this point. The expression for the interaction will contain the two interacting fields. Now the meson is described by a wave function which I shall call ϕ; the nucleon wave function shall be called ψ; and the so-called adjoint wave function of the nucleon shall be $\bar{\psi}$. The coupling with a pseudoscalar meson will further contain the Dirac operator known as γ_5—which is a very recondite thing. The two expressions for the coupling which have been most used are then:

(1) $G\bar{\psi}\gamma_5\psi\phi$ direct or pseudoscalar coupling

(2) $g\bar{\psi}\gamma_5\gamma_\mu\psi\dfrac{\partial\phi}{\partial x_\mu}$ pseudovector coupling

The first of these is known as "direct" coupling; it contains the meson wave function itself. The second type of coupling which was almost exclusively used in the literature until 1947 contains the derivative of the wave function of the meson with regard to the coordinates; it also contains some of the more ordinary Dirac operators, γ_μ where μ runs from 1 to 4. These two couplings are also known as the pseudoscalar and the pseudovector coupling.

Now the pseudovector coupling has one practical advantage, namely, that it permits theoretical physicists to operate with quantities with which they are more familiar. It has, however, a grave disadvantage, namely, that it gives a field theory with which you can calculate only the first order approximation to any process; the second order gives infinite results for any quantity you calculate. This is connected with the fact that the theory cannot be renormalized. The pseudoscalar interaction has the disadvantage that it uses the abstruse operator, γ_5, which has the strange property that it likes to change a particle of positive energy into a particle of negative energy or, in terms of the hole theory, that the most likely process is the formation of a pair of nucleons. This makes theoretical calculations somewhat more troublesome, but on the other hand, the pseudoscalar theory has the great advantage that it can be renormalized in the sense of field theory and gives finite results for every process in any approximation. So one of the questions is which of these theories is right, and, of course, the theorists hope that the pseudoscalar coupling is right.

The factors G and g in (1) and (2) are simply constants which determine the strength of the coupling. I have written the factor as capital "G" in (1) and as little "g" in (2) because the former is larger than the latter. A very important problem is clearly the determination of the coupling constant G or g; this plays the same role as the electric charge does in the interaction between charged particles and the electromagnetic field. This latter interaction is governed by the so-called fine-structure constant, $\frac{e^2}{\hbar c}$ which is 1/137 and, therefore, very small. On the other hand, the corresponding dimensionless quantity of meson theory, $\frac{G^2}{\hbar c}$, is about 15. This in contrast to $\frac{e^2}{\hbar c}$, is a large number, and this fact is the main cause of trouble in meson theory. All the methods which quantum mechanics has developed in the past were designed for small coupling between field and particle. This assumption of small coupling is very good for the electromagnetic field, and in this case we can predict effects of the order of one part in 10^9, and fit experiment, simply by making an expansion in powers of $\frac{e^2}{\hbar c}$. But it wouldn't be very successful to make an expansion in powers of $\frac{G^2}{\hbar c}$ which is 15: every successive order of approximation would give you a larger result than the previous one.

Now in exploring experimentally the coupling of mesons and nucleons, one turns to the simplest phenomenon which involves this interaction. This is the scattering of mesons by nucleons because it involves only one nucleon and one meson. A slightly more complicated phenomenon is the production of mesons by the interaction of electromagnetic radiation with nucleons, the photoproduction of mesons. In this case you have to consider, in addition to the meson and the nucleon, the interaction with the electromagnetic field. This interaction is known and simple. The next, more complicated, problem is that from which the theory started, namely the interaction between nucleons which is transmitted by the meson field. This is obviously more complicated because you now have to consider two nucleons and at least one meson. Finally, the most complicated of all phenomena that we hope to deal with is the production of mesons in the collision between two nucleons, in which case we not only have the mesons which produce the force between the two nucleons but, in addition, the meson which is produced in the process.

5. SCATTERING OF MESONS BY NUCLEONS

The scattering of mesons by nucleons has been investigated experimentally at various places, most extensively at the University of Chicago, with other important contributions by the teams at Columbia University and the University of Rochester. The results of these investigations seemed at first to be a major defeat for field theory because they seemed to agree with the pseudovector coupling. In making this comparison, the theoretical probability of scattering was calculated by the methods of perturbation theory, i.e., by using only the first term in a power series in $\frac{G^2}{\hbar c}$. In this approximation, pseudoscalar coupling gives a cross section for scattering of mesons by nucleons which is almost independent of energy, while pseudovector coupling yields a cross section which rises very rapidly with increasing energy. The experiments gave a very rapid increase of cross section with energy and thus favored the pseudovector coupling. However, the pseudovector coupling theory in perturbation (weak coupling) approximation predicted also some other things.

It predicted, for instance, that a negative meson interacting with a proton should in general be simply scattered. In principle, interaction between these two particles could also lead to the reaction $\pi^- + p = n + \pi^0$; (p = proton, n = neutron), but it was predicted by both pseudoscalar and pseudovector theories in the weak coupling approximation, that this reaction should not occur with sizeable probability. But experimentally, it does occur; in fact, the Chicago experiments show that it is about twice as powerful as the simple scattering of the negative mesons. Furthermore both theories in the weak coupling approximation predict that the positive-meson scattering should be about the same as the negative-meson scattering by protons. This again is wrong:

even if you add the charge-exchange scattering to the ordinary scattering of negative mesons, you still get only about a third of the probability of positive-meson scattering. The only possible conclusion from this is that the method of approximation is quite wrong, that one just can't get the right result by calculating merely the lowest power of G that occurs. If $\frac{G^2}{\hbar c}$ is 15, this isn't very surprising. The situation is best illustrated by a song by Arthur Roberts, an experimental physicist of the University of Rochester, who says,

> We have weak coupling and we have strong coupling,
> And we have wrong, as we knew all along, coupling.

A more reasonable procedure was proposed by Brueckner of Indiana University and his collaborators, Case and Watson. Taking the strong interaction seriously, they said that nucleon and meson can easily form a sort of compound; i.e., that there exists a virtual quantum state of the system of meson and nucleon. Postulating such a state, they could make use of a lot of calculations which had been made in the period from 1940–45 in an effort to explain the discrepancies between the observed properties of the μ-meson and the theory, and which are known as the strong coupling theory. This theory had predicted that there should be stationary "compound" states of nucleon and meson, and that the first of these should be a state of the following characteristics: the meson has an orbital momentum one, the whole system has a spin of three-halves so that the state can be described as a $P_{3/2}$ state, and the system is further characterized by a quantity which is called the isotopic spin which also has the magnitude three-halves, $T = 3/2$. I will not explain what that means.

Now, Brueckner proposed that the phenomena of meson scattering are governed by this excited state. The energy of the state is then experimentally determined to be about 300 million volts higher then the ground state of the nucleon. The scattering cross section, both for positive and negative mesons by protons, should then have a resonance maximum near the stationary state and should thus go essentially like this:

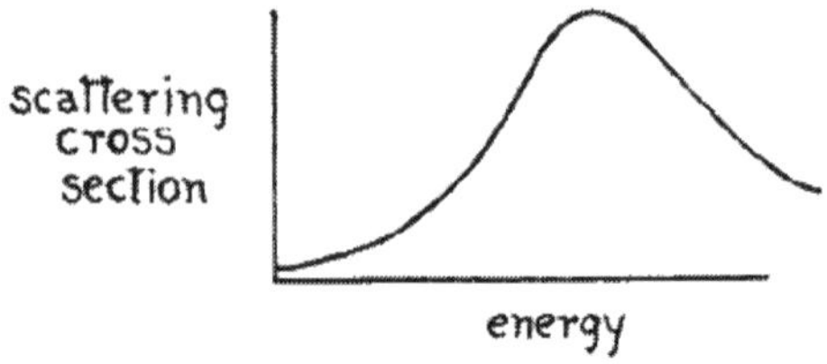

The maximum, of course, can be expected to be quite broad because there is a very strong probability that the stationary state will decay into a free meson and a nucleon, and large decay probability is equivalent to a large width of the state.

The main success of Brueckner's theory was that he could predict the ratio of the cross sections for positive-meson scattering and negative-meson scattering. The ratio of the total cross sections should be three to one, which is very close to the observed ratio. Brueckner and his co-workers could further predict that the charge-exchange scattering of negative mesons should be about twice as large as the ordinary scattering of negative mesons and that again agrees with experiment. They could further predict that the angular distribution of the scattered mesons should be about $1 + 3\cos^2\theta$—and this again gives a reasonable approximation to the observed distribution, although it is far from a complete description.

Scattering experiments are generally analyzed in terms of phase shifts of certain partial waves which describe the wave function of the particle. When this is done for meson scattering, it is found that the most important interaction is in the state $P_{3/2}$, $T = 3/2$. In addition to this there is also a strong interaction in the states of orbital momentum zero, that is in the S states, in spectroscopic notation, and that in this case there is strong interaction both for isotopic spin three-halves and for one-half. For three-halves there is strong repulsion; for one-half there is a somewhat weaker attraction.

6. THEORETICAL DEVELOPMENTS

Thus the Brueckner theory is quite successful, but, of course, it is purely phenomenological—the existence of an excited state is postulated but nothing is said about its origin. It is desirable to go back to fundamentals in order to explain this state. The road to this was opened by a young French physicist working at the Institute for Advanced Studies, Maurice Lévy, who developed a meson theory of nuclear forces about a year ago. So we are going back to the very beginning of the history of the subject, the theory of nuclear forces which gave the first lead on the pseudoscalar interaction. By consistent use of the pseudoscalar interaction Lévy was able to account for the phenomena observed in nuclear forces. The most important discovery which he made was that, as a direct consequence of pseudoscalar meson theory, the forces between two nucleons are strongly repulsive at small distances. This was the clue that had been missing in previous theories in which the two nucleons had always been considered as perfectly fixed in space. Theory then showed that two such nucleons would always have a strong attraction, indeed so strong that the two nucleons would fall into each other, and not form a stationary state of finite binding energy. Lévy's discovery saved the situation because he showed that there was at small distances a very strong repulsion which prevented the two nucleons from falling into each other.

After Lévy's calculations physicists began to wonder whether the pseudoscalar theory could throw any light on the meson-nucleon scattering experiments. As I said before, the attempt to account for these experiments by weak coupling theory had been a complete failure. The first success with direct application of pseudoscalar theory was achieved by Drell and Henley of Stanford University. They were able to show that between nucleon and meson the same kind of potential exists as between two nucleons according to Lévy, namely one which has a tremendously strong repulsion at small distances. At somewhat larger distances there is an attraction, mainly in the $P_{3/2}$ state. The strong repulsion is independent of angle and, therefore, acts primarily in states of zero angular momentum, that is in S states. Now if you have a strong repulsive potential and calculate the resulting cross section in the first Born approximation, you get a tremendously large result. Since in our case the repulsive potential acts in S states, the scattering is isotropic. It is easy to see that it should also be nearly independent of energy. All these results correspond exactly to the first-order theory which I previously described. The merit of Drell and Henley is that they showed exactly why the first-order result was wrong. Namely, if you have a potential which is strongly repulsive at small distances, and then maybe gets slightly attractive at larger distances, then

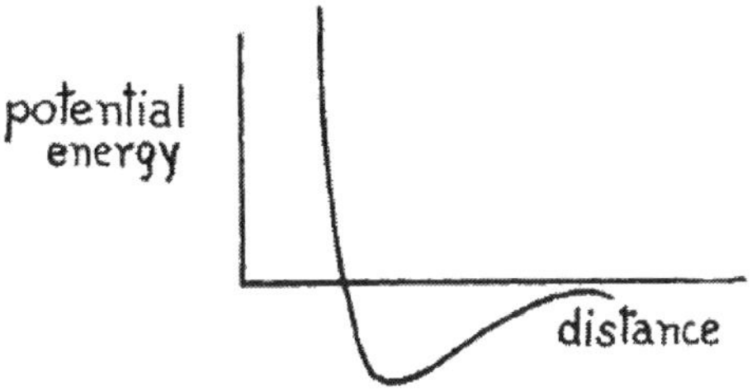

the only effect of the repulsion on the wave function is to make it essentially zero at the point where the repulsion stops. The most this can do is to give a phase shift proportional to the radius of the repulsive region, and this phase shift will be completely independent of the *magnitude* of the repulsive potential. So Drell and Henley showed not only that the weak coupling theory was wrong all along, but also *why* it was wrong and what should be done instead.

The next major progress was made by Chew, of the University of Illinois who did the same for the attraction that Drell and Henley had done for the repulsion; namely, he showed how one could calculate, at least in principle, the effects of the attractive force in a sensible way without using perturbation theory. He was able to show that for the $P_{3/2}$, $T = 3/2$ state one should indeed expect a resonance if one only makes suitable assumptions about the magnitude of the coupling constant. Chew used pseudovector coupling for convenience in calculation, but his theory can easily be translated into pseudoscalar coupling.

Building on all this work we, at Cornell University, started last fall to attack the problem from the beginning using the pseudoscalar interaction between nucleon and meson. We were able to explain qualitatively most of the features observed in the scattering experiments. In the first place, we get a strong repulsive interaction in the S state which gives an only moderately large S wave scattering, and this is just what the Chicago and Columbia experiments show. Then the theory gives an attraction in the $P_{3/2}$, $T = 3/2$ state, and the phase shift in this state can be adjusted to fit the experiments by proper choice of the coupling constant. There is just one unknown in the theory, namely, the coupling constant; if you fix the coupling constant to be about 15 you can explain quite well the observed phase shift including its dependence on energy. Finally, you find from the theory that the phase shifts for all the other P states are very small, and this again corresponds to observation. There is only one point which is not yet explained: experimentally, there is an attraction in the S state of isotopic spin $1/2$. But here we know that the theory is still deficient, because although we know that one should renormalize the theory, we are only now learning how to do the renormalization in practice.

7. CONCLUSION

I think that one can say at present that although the pseudoscalar meson theory is not yet able to explain quantitatively the meson-nucleon scattering, there is no cause for disbelieving it, because there is no qualitative discrepancy between the predictions of the theory and the experiments. It is likely to be just a matter of learning how to treat strong interactions before we can get quantitative results on meson scattering.

The question of nuclear forces, as I said, is much more complicated. Lévy's first attempt was extremely valuable because it showed that in principle the theory gave the right behavior of nuclear forces. In detail, numerous theoretical physicists have criticized Lévy's paper, and this is not surprising. However, the theory can explain why nuclei hold together, why you have strong forces, and why nucleons do not completely fall into each other. It predicts the interesting phenomenon of many-body forces; that is, it predicts that you have interactions not only between two nucleons, but also between three or more nucleons which hand a meson to each other around the circle. Weisskopf and his collaborators have pointed out that these many-body forces may be quite important for the explanation of the phenomenon of saturation of nuclear forces, that is, for the phenomenon that heavy nuclei also do not collapse. We can deduce from the pseudoscalar theory that nuclear forces depend on spin and deduce that there is a quadrupole moment of the deuteron.

A word must be said about other mesons. The mesons which I have talked about are the π-mesons which have a mass of about 300 electron masses. There are a lot of other mesons of much higher masses. At one time when people calculated only the first-order interaction, some physicists suggested that maybe these heavier mesons would prevent the collapse of nuclei. I think this was false. I think, in fact, that one can see now that these heavier mesons have little to do with the structure of nuclei. This is again connected with Lévy's potential which I mentioned before, which gives a strong repulsion between nucleons at distances of about 0.5×10^{-13} cm. Now the heavier mesons could only cause forces of shorter range than this; therefore, no matter what these forces are, they will get swamped by the strong repulsion which exists anyway because of the interaction of the nucleon with the π-meson. Therefore, I think one does not need to know much about heavier mesons in order to construct a satisfactory theory of nuclear forces. Of course, when two nucleons collide at very high speed, they can penetrate the region of mutual repulsion, and then heavy mesons can be produced (as we know experimentally) and probably influence the mechanism of the collision in an important way. But nuclear forces at moderate energies, for instance inside ordinary nuclei, appear to be transmitted mainly by π-mesons which are coupled to nucleons by pseudoscalar interaction. I believe it is only a matter of mathematical skill, but of very great mathematical skill, to extract from the theory the information which we know is buried in it.

"Ice-Cube – The Under-the-Ice Neutrino Observatory in Antarctica: Opening a New Window on the Universe from the South Pole"

Frances Halzen
Professor of Physics
University of Wisconsin-Madison

2,466th Meeting
November 4, 2022

Minutes of the 2466th Meeting

On November 4, 2022, from the Powell Auditorium of the Cosmos Club in Washington, D.C. and streaming online via Zoom and the PSW Science YouTube channel, President Larry Millstein called the 2,466th meeting of the Society to order at 8:05 p.m. EST. He announced the order of business and welcomed new members. The minutes of the previous meeting were then read and accepted.

President Millstein then introduced the speaker for the evening, Francis Halzen, Vilas Research and Gregory Breit Distinguished Professor of Physics at the University of Wisconsin–Madison and Principal Investigator of IceCube. His lecture was titled, "ICE-CUBE: The Under-the-Ice Neutrino Observatory in Antarctica, Opening a New Window on the Universe from the South Pole."

Halzen began by noting that theorist Wolfgang Pauli first proposed the existence of neutrinos in 1932, saying "I have done a terrible thing, I have invented a particle that cannot be detected." The neutrino's role, he said, was to "save the laws of physics." In the years following Pauli's hypothesis, scientists confirmed the existence of neutrinos via indirect evidence.

Neutrinos are detected in nuclear reactions and in nuclear reactors where they turn cooling water blue. Halzen said the phenomenon is created when a charged particle passes through water and emit blue light. Neutrinos are also emitted by cosmic rays bombarding Earth's atmosphere, making nuclear reactions with nitrogen and oxygen to release neutrinos and muons.

Viewing the night sky in infrared reveals the microwave background the universe. Going to blue wavelengths then reveals a narrow band across the sky, and increasing the blueness eventually results in a black sky image. Halzen said that blackness could be explored by detecting neutrinos. That is the research he has sought to conduct.

Unlike photons from the microwave background, neutrinos that reach Earth from the edge of the universe carry information about their origin; they are also difficult to detect. To detect neutrinos requires water and light sensors called photomultipliers. The idea of how to do so dates back to 1960, when Moisey Markov proposed to build deep in the clear ocean a cubic kilometer of light sensors pointed toward the sea floor. The idea was to detect neutrinos coming through the earth that collide with a nucleus in the water, emitting detectable blue light and muons from which neutrinos' trajectory can be calculated. Telescopes can then be directed to view the neutrino's origin and watch for an event that emits the neutrino.

Halzen realized it would be safer to place photomultipliers in ice, rather than in the open ocean where they would be exposed to an unfriendly environment. His idea eventually became IceCube, a neutrino detector installed near the South Pole where the ice is approximately three kilometers thick. He said that one-and-a-half kilometers below the surface, where the ice is 50,000 years old, it is free of bubbles and perfectly clear. There, blue light travels for hundreds of meters.

The speaker then described the IceCube construction and its instruments. The project was initially comprised of 500 photomultipliers measuring 10 inches each, set one and a half kilometers into the ice. The photomultipliers are organized into kilometer-long strings sunk deep into the ice. Today, IceCube is comprised of 5,160 photo-multipliers set into the Antarctic ice at depths between 1.4 and 2.4 kilometers below the surface. Halzen said once frozen, the detectors are completely safe from external harm or interruption, even through the SARS-CoV-2 pandemic. He then showed images and described the process his team used to bore holes through the ice and position the detectors.

IceCube detects 100 billion muons per year. Of those, 100,000 are from the atmosphere. From those atmospheric neutrinos, IceCube is searching for cosmic neutrinos. To do so, Halzen explained how the project measures neutrino energies and trajectories. He showed animated data to illustrate what it looks like when IceCube detects a cosmic neutrino. It detects approximately 200 such neutrinos every year.

IceCube's conclusions thus far are that it detects the Universe rather than our own galaxy. In the extreme universe more energy is emitted in neutrinos than gamma rays, and those conclusions represent an opportunity for a "new" astronomy. When IceCube detects a cosmic neutrino, the team forwards that data to telescopes around the world to give them opportunity to look at the neutrino's point of origin. In September 2017, the MASTER robotic optical telescope network associated a detected neutrino with a flaring blazar from an active galaxy within the Orion constellation.

Earlier on the day of the lecture, Science published Halzen's conclusion that IceCube accurately detected Galaxy NGC 1068 as a neutrino source. He explained in depth the process his team used to interpret their data.

After he concluded his lecture, the speaker answered questions from the in-person and online viewing audiences. One member asked about the theoretical limit of detector resolution. The speaker said the key is to understand the optics of Antarctic ice, which is an ongoing process. He said glaciology research is a collateral research area for the project. Another member asked about cost. The speaker said IceCube cost about $300 million, and has a proposal to construct a detector eight times as large for the same amount of money.

After the question and answer period, President Millstein thanked the speakers, made the usual housekeeping announcements, and invited guests to join the Society. President Millstein then adjourned the meeting.

Temperature in Washington, D.C.: 18° C

Weather: Partly Cloudy
Attendance: 47 in-person and 40 streaming for a total of 87 live viewers. The recording was viewed 459 times in the first two weeks of posting.

Respectfully submitted,
Recording Secretary James Heelan

THE TELESCOPES

"On the Progress of the Construction of the New Telescope for the Naval Observatory"
Simon Newcomb
United States Naval Observatory

30th Meeting
June 29, 1872

Simon Newcomb reported on the construction of the 26-inch "Great Equatorial" telescope. Within two years, it will be used by Asaph Hall to discover the two moons of Mars, Phobos and Deimos.

30th Meeting. 1874. *Bulletin of the Philosophical Society of Washington* (1): 62.

"Building the Giant Magellan Telescope"
James Fanson
Project Manager
GMTO Corporation

2,436th Meeting
March 5, 2021
https://pswscience.org/meeting/2436/

30th Meeting. June 29, 1872.

Tbe President in the Chair.

Mr. S. Newcomb made a verbal communication

ON THE PROGRESS OF THE CONSTRUCTION OF THE NEW TELESCOPE FOR THE NAVAL OBSERVATORY.

(*The substance of these remarks is published in Scribner^s Monthly Magazine^ October^* 1873.)

Mr. W. B. Taylor made some verbal comments on a recent publication

ON OUR PRESENT KNOWLEDGE OP THE PLANET JUPITER.

"Building the Giant Magellan Telescope"
James Fanson
Project Manager
GMTO Corporation

2,436th Meeting
March 5, 2021

Minutes of 2,436th Meeting
On March 5, 2021, by Zoom videoconference broadcast on the PSW Science YouTube channel, President Larry Millstein called the 2,436th meeting of the Society to order at 8:02 p.m. EST. He announced the order of business and welcomed new members. The Recording Secretary then read the minutes of the previous meeting.

President Millstein then introduced the speaker for the evening, James Fanson, Project Manager for the Giant Magellan Telescope. His lecture was titled, "Building the Giant Magellan Telescope."

Franson began by saying the human eye has an aperture of approximately 0.2 cm^2 and requires assistance to see further than the Andromeda Galaxy, approximately 2.5 million light-years away. Since Galileo's telescope in 1609, the aperture diameter of telescopes has doubled approximately every 30 years. Galileo used a 2.6 cm lens in his telescope. Almost three hundred years later, in 1897, the Yerkes telescope reached the practical limits of refracting telescopes with its 1 meter lens.

The next generation of telescopes used mirrors to better focus light and improve image resolution at further distances. Built in 1934, the Palomar telescope reached a monolithic mirror size limit of 5 meters. In 1990, the Keck telescope broke that barrier with its 10 meter segmented mirror.

Now that it has been 31 years since the Keck telescope, Fanson said it is time for the next doubling in telescope aperture. The Giant Magellan Telescope (GMT), the 30 m Telescope, and the European

Extremely Large Telescope (ELT) are all underway to achieve that feat.

The GMT will have a segmented mirror with computer control. The primary mirror will be composed of seven individual mirror segments, each 8.4 m in diameter. Computer control will allow the GMT to maximize image resolution by phasing the seven mirror segments and correcting for atmospheric distortions with adaptive optics. Fanson said these advancements will allow the GMT to get near the level of imaging resolution that has so far only been available from space-based telescopes.

The GMT will be constructed at the Las Campanas Observatory Site at the southern end of the Atacama Desert in Chile. The remote mountaintop location means it has minimal light pollution, few clouds, and dry, steady air. The GMT will be housed in an enclosure roughly the size of a football field, with its operating machinery located in a separate building. Fanson showed an animation and gave viewers a "tour" of the telescope, highlighting its mechanical functions.

Most modern telescopes are reflecting telescopes with primary and secondary mirrors. In the GMT, both the primary and secondary mirrors will be segmented, operating in pairs, arranged in the Gregorian optical design. These features together will be the first of their kind.

Fanson then showed and described the process by which the mirrors are created, using 20 tons of E6 borosilicate glass melted in a spinning furnace. Five of the seven primary mirror segments have so far been cast and polished to a precision of approximately 1/100,000,000 of an inch.

Once the mirrors arrive at their destination in Chile, they will receive a 200 nanometer thick aluminum coating and will be installed on the telescope structure. Each mirror will be affixed to 165 pneumatic actuators to allow alignment.

GMT will phase the primary mirror segments by using light from a guide star, reflected off the adjacent edges of adjacent segments. Based on those dispersed reflections, GMT will identify the misalignment of the edges by the degree to which the interference fringes have rotated. A prototype of the phasing system has been in use on the Magellan Clay telescope at the Las Campanas site.

Fanson then described the history of adaptive optics, first proposed by Horace Babcock in 1953, by which scientists can compensate for atmospheric distortion with actuators that can restore the reflected light to their original wavelength. Adaptive optics on the GMT will be accomplished by its segmented secondary mirror. Each segment will contain 672 actuators operating at 2 kHz bandwidth, allowing real-time wavefront correction.

Once completed, GMT will have 11 instrument stations. Among them will be G-CLEF, an optical echelle instrument that scientists will use to detect exoplanets, and two visible light spectrographs that can be fiber fed to widen their field of view.

Fanson described the engineering techniques GMT will use to protect it from seismic activity and wind vibrations. He described the current state of construction and operations. He then answered questions from the online viewing audience.

After the question and answer period, President Millstein thanked the speaker, made the usual housekeeping announcements, and invited guests to join the Society. At 9:56 p.m., President Millstein adjourned the meeting.

Temperature in Washington, D.C.: 4.5° C
Weather: Clear
Concurrent Viewers of the Zoom and YouTube live stream, 118 and views of the recoding on the PSW Science YouTube and Vimeo channels in the first two weeks: 393.

Respectfully submitted,
James Heelan, Recording Secretary

PLANETARY BODIES THEN AND NOW

Hall Discovers the Moons of Mars

"The Results of His Search for Satellites of Mars"
Asaph Hall
Washington Observatory

136th Meeting
February 2, 1878

Asaph Hall announces the discovery of Phobos and Deimos, the two moons of Mars, one of the most important astronomical discoveries of the 19th century.

136th Meeting. 1874. *Bulletin of the Philosophical Society of Washington* (2): 186-187.

Stern Discovers How Interesting Pluto Is

"The Exploration of Pluto"
Alan Stern
Principal Scientist, New Horizons
Vice President for Research, Southwest Research Institute

2,362nd Meeting
April 15, 2016
https://pswscience.org/meeting/the-exploration-of-pluto/

137TH MEETING. FEBRUARY 16, 1878.

Vice-President WELLING in the Chair.

Thirty-eight members and visitors present.

Mr. PARKER, for a Committee appointed at the last meeting reported the following resolutions, which were unanimously adopted :—

The Philosophical Society of Washington having heard with unusual interest the communication of Professor ASAPH HALL, U. S. N., giving an account of his discovery of two satellites of Mars, therefore,

Resolved, That this important discovery by one of its members constitutes an event which not only the Society appreciates, but which will also be regarded with interest by the present age, so distinguished for its steady advance into the realms of the unknown.

While the members of the Society congratulate Professor HALL upon the success which has crowned his persevering efforts, and admire the modesty which characterizes his account of them, they feel proud to claim him as one of their number.

As every one who makes an important discovery either in the arts or sciences, beneficial to mankind, stamps a coin that will transmit his name with honor to future ages, so in time to come will be perpetuated the name of the distinguished discoverer of "Δείμος" and "Φόβος".

Resolved, That a copy of the above resolution, engrossed and signed by the President and Secretaries, be presented to Professor HALL, with the request that he will permit his paper to be published in the Bulletin of the Society.

Mr. LESTER WARD continued a communication on

THE NATURAL SYSTEM OF PLANTS.

Mr. ANTISELL made remarks on the proper basis of classification; Mr. WHITE on fossil plants; and Mr. GILL on the cautiousness with which resemblances should be used, though they may be taken as a guide in first examinations, giving instances of strong resemblances in animals of different orders, and of superficial resemblance and different anatomical structure.

Mr. G. K. GILBERT made a communication on

THE RECENT HISTORY OF THE GREAT SALT LAKE,

giving the evidence, obtained while engaged in investigating the agricultural resources of Utah, of an increase in the flow of the streams, and rise and increased area of the lake itself. There were also annual fluctuations caused by the melting snows in spring and the heat in summer. He also discussed the causes to which the progressive rise has been attributed, viz., volcanic action, change of climate, agency of man. He spoke also of similar increase in the streams and lakes of Colorado.

Remarks were made by Mr. HAYDEN and Mr WHITE.

"The Exploration of Pluto"
Alan Stern
Principal Scientist, New Horizons
Vice President for Research, Southwest Research Institute

2,362nd Meeting
April 15, 2016

President Larry Millstein called the 2362nd meeting of the Society to order at 8:04 p.m. He announced the order of business and welcomed new members. The minutes of the previous meeting were read and approved. President Millstein summarized the proceedings of the 26th meeting of the Society in 1872. President Millstein then introduced the speaker for the evening, Alan Stern, Principal Scientist for the New Horizons Mission, Vice President for Research at the Southwest Research Institute, and CEO of Uwingu and The Golden Spike Company. His lecture was titled "The Exploration of Pluto".

Stern began by explaining that New Horizons was the fastest spacecraft ever launched, flown 3 billion miles and 3000 days to the farthest world humans have ever explored. The United States has been first to every planet in our solar system, beginning with the flyby of Venus by Mariner 2 in 1962. The surveying of Pluto by New Horizons in 2015 completed this 53 year project and began a new phase in our exploration of the solar system.

New Horizons cost approximately 700 million dollars over its 15 year initial mission. Stern observed that such an enormous investment requires demonstrating to the scientific community that the program is a funding priority. The New Horizons team made their case using various observations, including infrared reflectivity surveys identifying molecules such as carbon dioxide, methane, and nitrogen moving on the surface—a richer surface environment than Mars. Atmospheric refraction indicated that Pluto had an atmosphere. The team also showed that Pluto's moon Charon was formed through a collision, closely analogous to the formation of Earth's moon.

Perhaps the principal factor in finally elevating New Horizons to the top of the funding priority list was a paradigm shift in the 1990s. The previous conventional wisdom was that the solar system consisted of four small rocky inner planets close to the sun, with four gas giants beyond them. Pluto did not fit this pattern. Advances in technology, however, began to identify further objects beyond Pluto. Instead of being a misfit, Pluto was simply the first of a new trend, the Kuiper Belt objects. The Kuiper Belt is now the single largest structure in the solar system, and its "dwarf planets" became a third class of planets—and by far the most numerous. Thus, the Pluto mission became a voyage toward truly new horizons.

The spacecraft designed for the mission was small, powered, appropriately enough, by Plutonium, because it would be too far away from the sun to use solar panels. It carries seven scientific instruments that are so miniaturized that all of them together weigh less than the camera aboard the Cassini mission that imaged Saturn in 2004.

The other advance of New Horizons was its massive launch vehicle, the Atlas rocket. Loaded with only the compact New Horizons probe, the rocket was launched essentially empty, providing maximum speed. The rocket broke the sound barrier going vertically, reaching 18,000 MPH in Earth orbit, and, ultimately 50,000 MPH, resulting in 14 Gs of acceleration. The Apollo mission, at 25,000 MPH, required three days to reach the moon. New Horizons passed the moon in nine hours.

The Pluto flyby was completed successfully in 2015, but it will take 16 months to download all 15 gigabits of data back to Earth. The data we have received thus far is already changing the way we see Pluto. Images from the far side of Pluto show layers of haze, as the atmosphere is forced over mountains. Pluto's sky is blue like Earth's because tiny particles in the atmosphere scatter the sun. New Horizons also observed 1,000 foot tall crystals of methane, and "cryovolcanoes" erupting ice. One of the biggest mysteries to be investigated using these new data is why Pluto is still geologically active. Even as the Pluto data is slowly transmitted back to Earth, the New Horizons mission continues toward the Kuiper Belt to image several large objects already identified for surveying.

Stern concluded by observing that a small robot millions of miles away captivated and inspired people all over the world, and he urged us to keep exploring.

After the conclusion of the talk, President Millstein invited questions from the audience.

One questioner asked about the difficulty of hitting such a small target. Stern admitted that when New Horizons launched, Pluto's orbit was not known well enough to complete the mission. Scientists pooled all available data to fine tune the knowledge of Pluto's location and speed, going as far as re-analyzing the original Clyde Tombaugh images from the 1930s using modern data analysis tools to get the longest possible timespan.

Another questioner asked what would happen if New Horizons hit any debris, given its tremendous speed. Stern explained that the Pioneer 10 and 11 and Voyager 1 and 2 missions provided an opportunity to characterize the interplanetary dust environment. Based on these findings, multilayer insulation with Kevlar was added to mitigate the risk from dust, but a participle the size of a grain of rice would still shred the spacecraft. Thus, the team actively imaged the route looking for rings and dust sheets, prepared to divert the probe at the first sign of such hazards. Ultimately, Stern explained, exploring requires taking risks.

After the question and answer period, President Millstein thanked the speaker, made the usual housekeeping announcements, and invited guests to join the Society. At 10:09 p.m., President Millstein adjourned the 2,362nd meeting of the Society to the social hour.

Attendance: 143
The weather: Clear
The temperature: 11°C
Respectfully submitted,

Preston Thomas
External Communications Director

Made in United States
Orlando, FL
09 November 2026

90443352R00114